AS/A-LEVEL YEAR 1

STUDENT GUIDE

AQA

Biology

Topics 3 and 4

Organisms exchange substances with their environment

Genetic information, variation and relationships between organisms

Pauline Lowrie

PHILIP ALLAN FOR HODDER EDUCATION
AN HACHETTE UK COMPANY

Philip Allan, an imprint of Hodder Education, an Hachette UK company, Blenheim Court, George Street, Banbury, Oxfordshire OX16 5BH

Orders

Bookpoint Ltd, 130 Milton Park, Abingdon, Oxfordshire OX14 4SB

tel: 01235 827827

fax: 01235 400401

e-mail: education@bookpoint.co.uk

Lines are open 9.00 a.m.–5.00 p.m., Monday to Saturday, with a 24-hour message answering service. You can also order through the Hodder Education website: www.hoddereducation.co.uk

ISBN 978-1-4718-4357-0

First printed 2015

Impression number 5 4 3 2 1

Year 2018 2017 2016 2015

Cover photo: Sergey Nivens/Fotolia

Typeset by Integra Software Services Pvt Ltd, Pondicherry, India

Printed in Italy

Contents

Content Guidance

Questions & Answers

Getting the most from this book

Exam tips

Advice on key points in the text to help you learn and recall content, avoid pitfalls, and polish your exam technique in order to boost your grade.

Knowledge check

Rapid-fire questions throughout the Content Guidance section to check your understanding.

Knowledge check answers

1 Turn to the back of the book for the Knowledge check answers.

Summaries

- Each core topic is rounded off by a bullet-list summary for quick-check reference of what you need to know.

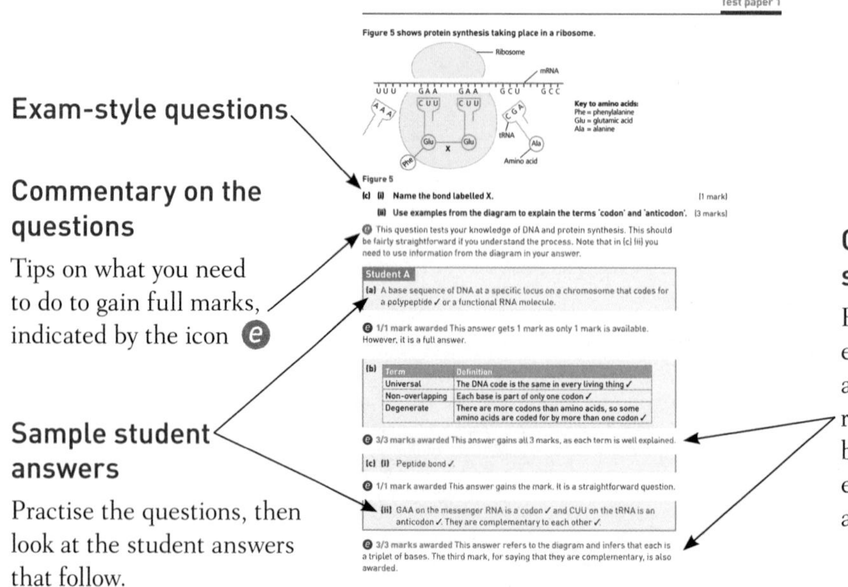

Test paper 1

Figure 5 shows protein synthesis taking place in a ribosome.

Key to amino acids:
Phe = phenylalanine
Glu = glutamic acid
Ala = alanine

Figure 5

(c) (i) Name the bond labelled X. (1 mark)

(ii) Use examples from the diagram to explain the terms 'codon' and 'anticodon'. (3 marks)

e This question tests your knowledge of DNA and protein synthesis. This should be fairly straightforward if you understand the process. Note that in (c) (ii) you need to use information from the diagram in your answer.

Student A

(a) A base sequence of DNA at a specific locus on a chromosome that codes for a polypeptide ✓ or a functional RNA molecule.

e 1/1 mark awarded This answer gets 1 mark as only 1 mark is available. However, it is a full answer.

(b)

Term	Definition
Universal	The DNA code is the same in every living thing ✓
Non-overlapping	Each base is part of only one codon ✓
Degenerate	There are more codons than amino acids, so some amino acids are coded for by more than one codon ✓

e 3/3 marks awarded This answer gains all 3 marks, as each term is well explained.

(c) (i) Peptide bond ✓

e 1/1 mark awarded This answer gains the mark. It is a straightforward question.

(ii) GAA on the messenger RNA is a codon ✓ and CUU on the tRNA is an anticodon ✓. They are complementary to each other ✓.

e 3/3 marks awarded This answer refers to the diagram and infers that each is a triplet of bases. The third mark, for saying that they are complementary, is also awarded.

Exchange (Topic 3) • Genetics (Topic 4) 61

Commentary on sample student answers

Find out how many marks each answer would be awarded in the exam and then read the comments (preceded by the icon e), which show exactly how and where marks are gained or lost.

About this book

This guide will help you to prepare for AQA AS/A-Level Biology Year 1, topics 3 and 4. At AS, topics 3 and 4 form half the content of papers 1 and 2. At A-level these topics are examined in paper 1 (together with topics 1 and 2) and in paper 3 (together with topics 1, 2 and 5–8).

The **Content Guidance** covers all the facts you need to know and concepts you need to understand for topics 3 and 4. It is really important that you focus on *understanding* and not just learning facts, as the examiners will be testing your ability to apply what you have learned in new contexts. This is impossible to do unless you really understand everything. The Content Guidance also includes examiner tips and knowledge checks to help you prepare for your exams.

The **Questions & Answers** section shows you the type of questions you can expect in the exam. It would be impossible to give examples of every kind of question in one book, but these should give you a flavour of what to expect. Two students, student A and student B, have attempted each question. Their answers, along with the accompanying comments, should help you to see what you need to do to score a good mark — and how you can easily *not* score a mark even though you probably understand the biology.

What can I assume about the guide?

You can assume that:

- the basic facts you need to know and understand are stated explicitly
- the major concepts you need to understand are explained clearly
- the questions at the end of the guide are similar in style to those that will appear in the final examination
- the questions assess the different assessment objectives
- the standard of the marking is broadly equivalent to that which will be applied to your answers

How should I use this guide?

This guide lends itself to a number of uses throughout your course — it is not *just* a revision aid. You could:

- use it to check that your notes cover the material required by the specification
- use it to identify your strengths and weaknesses
- use it as a reference for homework and internal tests
- use it during your revision to prepare 'bite-sized' chunks of related material, rather than being faced with a file full of notes

You could use the Questions & Answers section to:

- identify the terms used by examiners and show what they expect of you
- familiarise yourself with the style of questions you can expect
- identify the ways in which students gain, or fail to gain, marks

Develop your examination strategy

Just as reading the *Highway Code* alone will not help you to pass your driving test, this guide cannot help to make you a good examination candidate unless you develop and maintain all the skills that examiners will test in the final exams. You also need to be aware of the type of questions examiners ask and where to find them in the exams. You can then develop your own personal examination strategy. But, be warned, this is a highly personal and long-term process; you cannot do it a few days before the exam.

Things you *must* do

- Clearly, you must know some biology. If you don't, you cannot expect to get a good grade. This guide provides a succinct summary of the biology you must know.
- Be aware of the skills that examiners *must* test in the exams. These are called assessment objectives and are described in the AQA Biology specification.
- Understand the weighting of the assessment objectives that will be used. These are as follows.

Assessment objective	Brief summary	Marks in A-level paper 1/%	Marks in A-level paper 2/%	Marks in A-level paper 3/%	Marks in AS paper 1/%	Marks in AS paper 2/%
AO1	Knowledge and understanding	44–48	23–27	28–32	47–51	33–37
AO2	Application of knowledge and understanding	30–34	52–56	35–39	35–39	41–45
AO3	Analyse, interpret and evaluate scientific information, ideas and evidence	20–24	19–23	31–35	13–17	21–25

- Use past questions and other exercises to develop all the skills that examiners must test. Once you have developed them all, keep practising to maintain them.
- Understand where in your exams different types of questions occur. For example, the final question on AS paper 2 will always be worth 10 marks and will test AO1 by requiring you to write extended prose. If that is the skill in which you feel most comfortable, and many AS students do, why not attempt this question first?
- Remember that mathematical skills account for about 10% of the marks. Do make sure you can carry out these calculations, including percentages, ratios and rates of reaction.
- You need to be familiar with the techniques you have learned in the required practicals, and be able to describe how these techniques might be used in a different context. Also, you need to be able to evaluate practical investigations and data presented to you in the exam. Answers to the questions set in the required practical are given on page 91.

Content Guidance

■ Organisms exchange substances with their environment

Surface area to volume ratio

A unicellular (single-celled) organism can take in molecules all over its surface. However, in a multicellular organism, the cells on the inside cannot exchange materials with their environment, as shown in Figure 1.

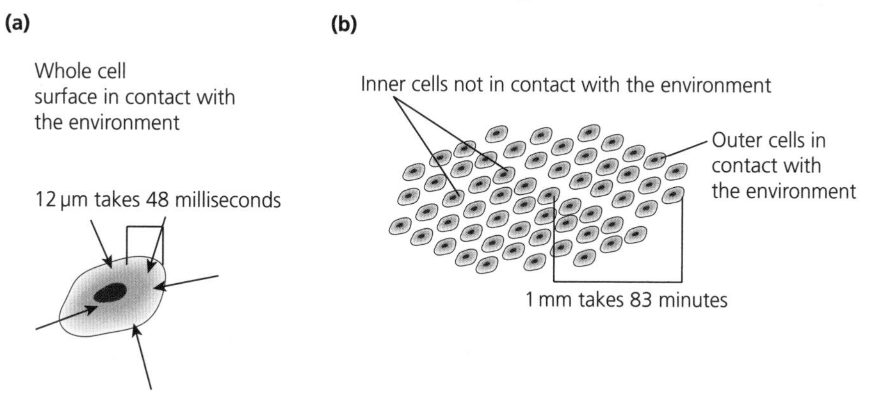

Figure 1 Diffusion in (a) unicellular and (b) multicellular organisms

You can see that the cells in the middle of a multicellular organism are a long way from the outside. It would take far too long for diffusion to supply molecules to these cells.

The rate of diffusion between a multicellular organism and its environment depends on:

- the total surface area of the organism in contact with the environment
- the total volume occupied by all the cells in the organism

As the number of cells in an organism increases, the volume of the organism increases. The organism's surface area also increases, but not by so much because some of the cells are not in contact with the environment. Therefore, it is important to consider the proportion of surface area to volume, or the surface area to volume ratio.

Small multicellular organisms exchange molecules all over their body surface if their body shape is adapted. Figure 2 shows a flatworm that lives in freshwater. These are about 5–25 mm long. You can see that the flatworm has a thin, flat shape. This means that all the cells in the flatworm are close to the environment, so it can rely on diffusion to supply all its cells.

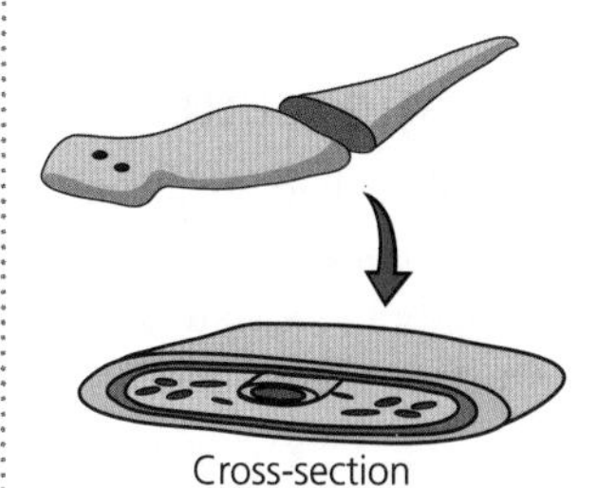

Figure 2 A flatworm

Organisms that are larger have to rely on specialised exchange surfaces, such as the lungs in a human where gas exchange takes place. An efficient gas exchange surface has:

- a large surface area
- a thin exchange surface, i.e. a short distance for the molecules to travel
- a steep diffusion gradient so that diffusion per unit time is greater

Another factor that affects the surface area to volume ratio is metabolic rate. Organisms with a high metabolic rate respire at a faster rate, so they need more oxygen to diffuse in and more carbon dioxide to diffuse out per unit time.

Organisms with a high metabolic rate generate more heat as this is a by-product of respiration. In a cold environment, large organisms lose less heat and so are at an advantage. However, in a warm environment they may have difficulty in losing enough heat.

Knowledge check 1

Newly hatched tadpoles have large feathery external gills that are well supplied with capillaries. Explain the advantage of this.

Knowledge check 2

Which features of the epithelial cell from the small intestine give it:

a a large surface area, and

b a short diffusion distance?

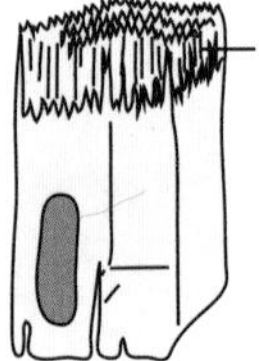

Exam tip

Make sure you don't just refer to 'surface area' alone. Larger organisms have a larger surface area than smaller organisms, but it is the surface area to volume ratio that is important.

Summary

- As an organism increases in size, its surface area to volume ratio decreases.
- Some multicellular organisms have adaptations to their body shape that increase their surface area to volume ratio.
- Organisms with a higher metabolic rate generate more heat as a by-product of respiration. This gives them an advantage at low environmental temperatures but they may have difficulty losing heat over their body surface when environmental temperatures are high.

Gas exchange

Single-celled organisms

Single-celled organisms can exchange gases across the whole of their cell-surface membrane. They need no special adaptations as the whole of their cell-surface membrane is in touch with the environment. Their exchange surface is the cell-surface membrane.

Insects

Insects are multicellular and they have an external skeleton that is coated with a waxy substance, so they cannot exchange gases across their body surface. Air enters their body through openings called **spiracles** that lead to a system of air-filled tubes called **tracheae** and **tracheoles**, as shown in Figure 3.

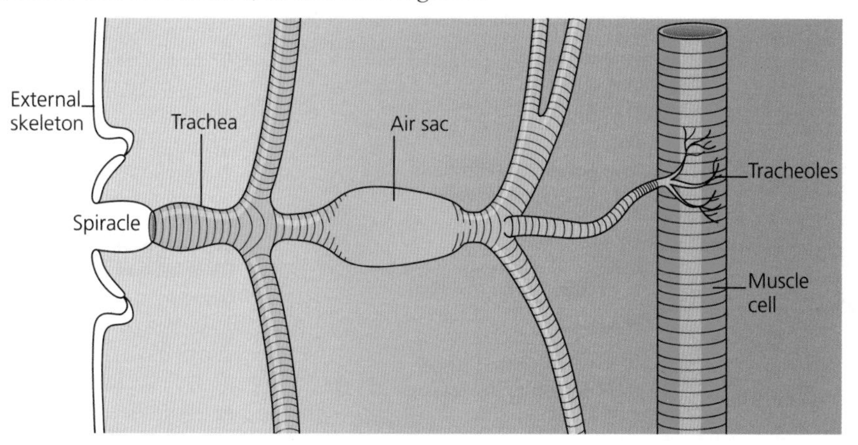

Figure 3 The tracheal system of an insect

The tracheae have rigid rings made of chitin that hold them open. The tracheae branch into smaller tracheoles that penetrate between the cells and right into the muscle fibres. It is here that gas exchange takes place. In some insects, diffusion alone supplies enough oxygen to the muscle cells. However, larger insects ventilate the tracheal system by closing and opening the spiracles and contracting abdominal muscles which squeeze the tracheal system. This pumps air through the system.

Knowledge check 3

What is the gas exchange surface in an insect?

Fish

Fish obtain their oxygen from the water they swim in. Water passes over their gills, which are adapted to absorb oxygen from the water. You can see the structure of a gill in Figure 4.

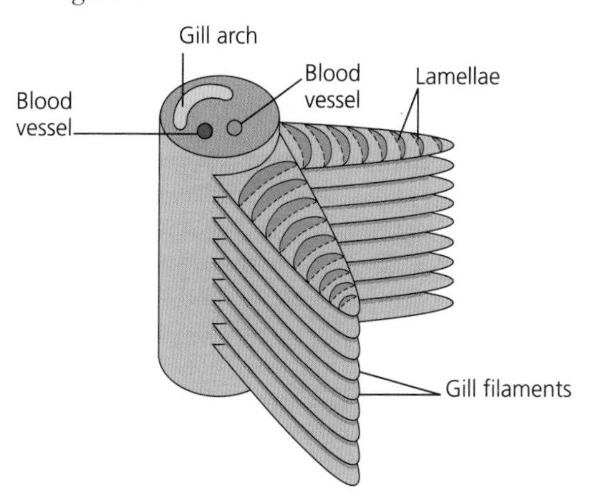

Figure 4 The structure of a fish gill

The gill arch is bony and it supports two stacks of thin **filaments** that have rows of thin **lamellae** over their surface. The lamellae stick up vertically from the filaments. The surface of the lamellae consists of a single layer of thin, flattened cells. Underneath this layer is a network of capillaries.

The blood vessels in the lamellae are arranged so that water and blood flow in opposite directions. This is called the counter-current system (Figure 5).

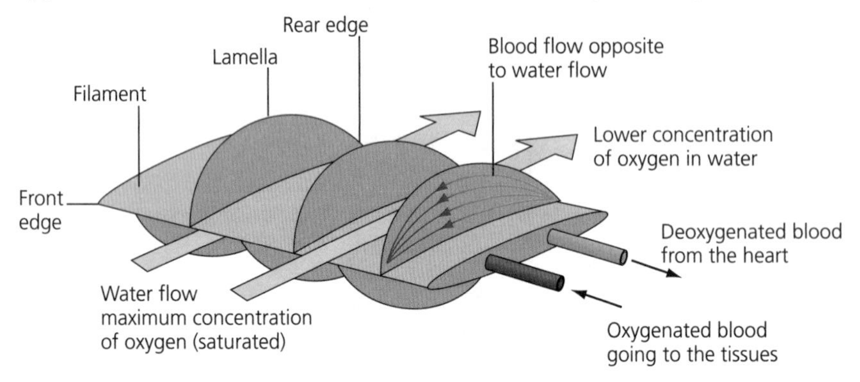

Figure 5 The counter-current flow of water over a gill filament

The advantage of a counter-current system is that it maintains a concentration gradient all the way along the capillary, as shown in Figure 6.

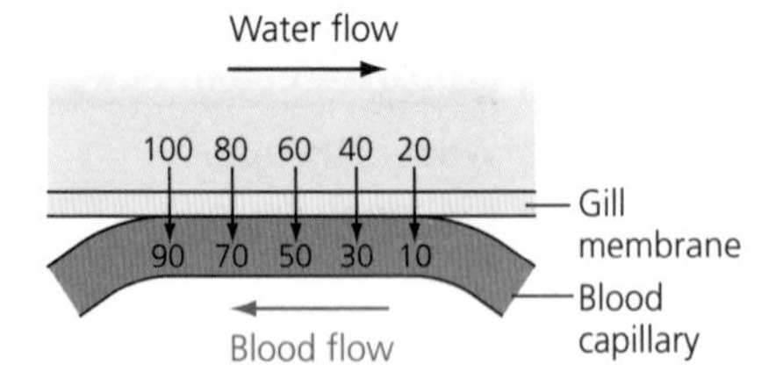

Figure 6 In the counter-current system, the gradient of oxygen concentration is maintained between water and blood. The figures show the % saturation with oxygen

Knowledge check 4

What would happen if water flow and blood flow were in the same direction?

Knowledge check 5

What features of the gas exchange system in a fish provide:

- **a** a large surface area
- **b** a short diffusion pathway, and
- **c** a large concentration gradient?

Dicotyledonous plants

Gas exchange in dicotyledonous plants takes place inside the leaves. You can see the structure of a leaf in Figure 7.

The main function of a leaf is to carry out photosynthesis, and for this it needs to take in carbon dioxide and water. The mesophyll cells that contain chloroplasts need plenty of sunlight, which is why the leaves are broad and thin. However, to obtain enough carbon dioxide they also need a large exchange surface. The gas exchange surface in a leaf is the surface of the mesophyll cells. In Figure 7, it is easy to see that the spongy mesophyll cells have most of their surface exposed to the air spaces in the leaf. You should also see that the palisade mesophyll cells have small gaps between them, again making sure that most of their surface is in contact with the air.

Knowledge check 6

How is gas exchange in a leaf different at night?

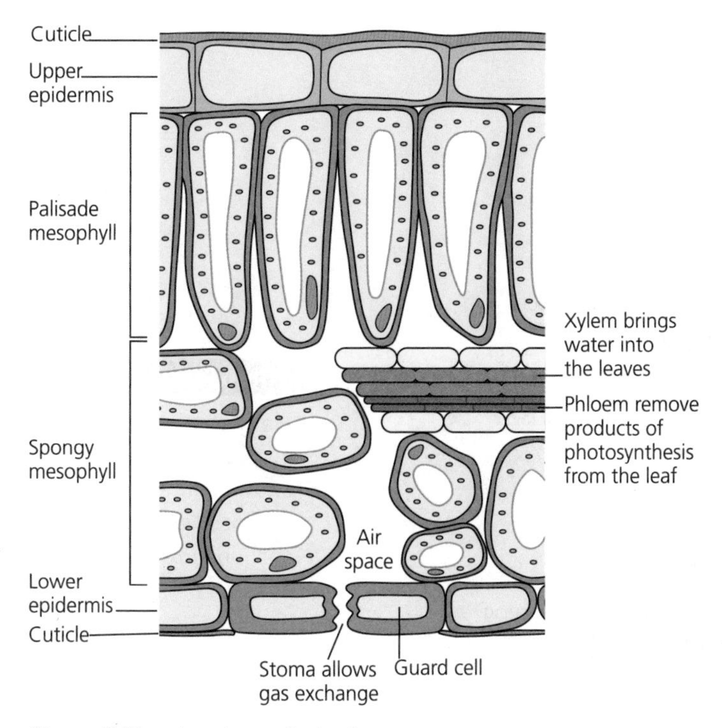

Figure 7 The structure of a leaf

Air enters the leaf through the stomata. By day, this air is higher in carbon dioxide than the cytoplasm of the cells. Therefore, carbon dioxide diffuses into the cells. The mesophyll cells are higher in oxygen than the air spaces, so oxygen diffuses out of the cells into the air spaces. The oxygen then diffuses out of the leaf via the stomata.

Exam tip

Remember that plants respire continuously but only photosynthesise in sunlight.

Preventing drying out

Organisms that live on land need to avoid drying out. As their gas exchange systems have a large surface area, a great deal of water could be lost.

Insects reduce the chances of drying out by having:

- an outside skeleton that is rigid and coated with a waxy substance, making it waterproof
- spiracles that can be closed using valves (in some insects)
- hairs surrounding the spiracles to trap humid air and reduce the water potential gradient between the tracheae and the outside air

Leaves are adapted to reduce water loss by having:

- a waterproof cuticle on the upper and lower surfaces of the leaf
- stomata that can be closed if water loss is too high

However, some plants, called **xerophytes**, have additional adaptations to reduce water loss. This is because they are adapted to grow in dry conditions. One example of a xerophyte is a cactus (Figure 8).

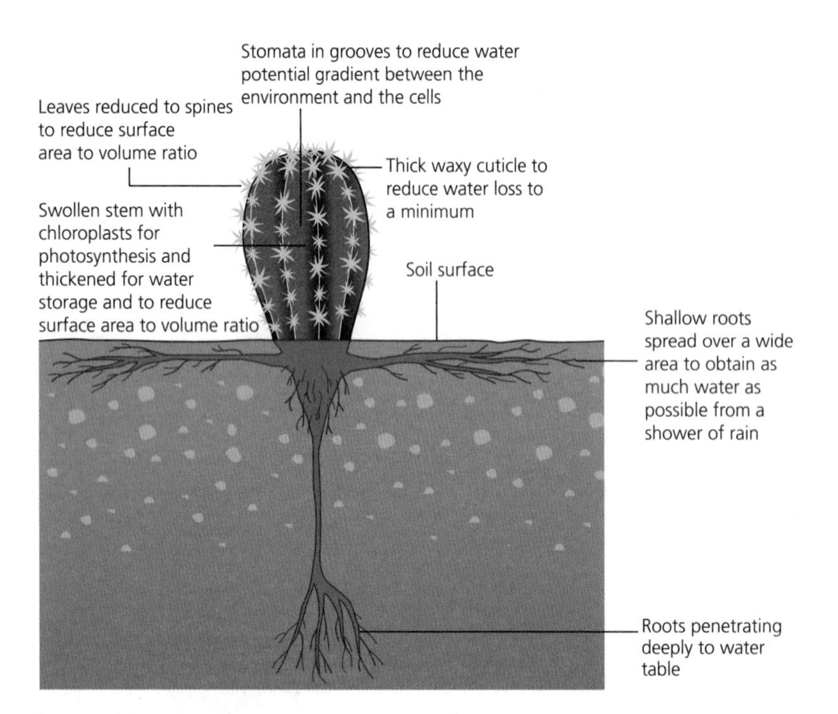

Figure 8 The features of a cactus that help it to survive in a dry habitat

Another xerophyte is marram grass, which lives in sand dunes. You can see a cross-section diagram of its leaf in Figure 9.

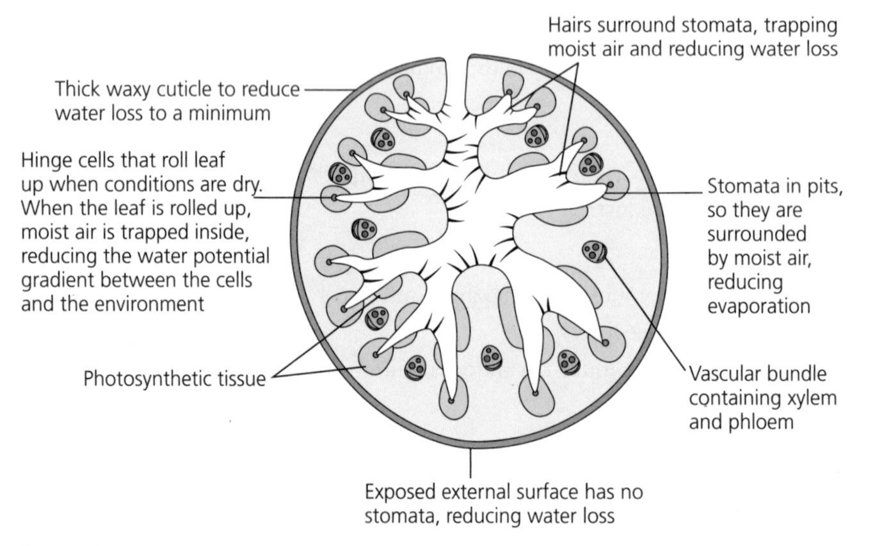

Figure 9 Marram grass is well adapted to restrict water loss

Knowledge check 7

Many cacti use a special kind of photosynthesis, which means they can take in carbon dioxide at night and store it. Explain how this helps them to control their water loss.

Exam tip

When you are asked for xerophytic adaptations, don't just describe the features — explain how they reduce water loss.

Exam tip

Sometimes you are given a diagram or photograph of a xerophyte and asked for a feature *visible in the diagram or photograph*. Make sure that you can apply your understanding to specific examples.

Gas exchange in humans

In humans, the gas exchange surface is the alveoli inside the lungs. The lungs are in the chest cavity. The structure of the human breathing system is shown in Figure 10.

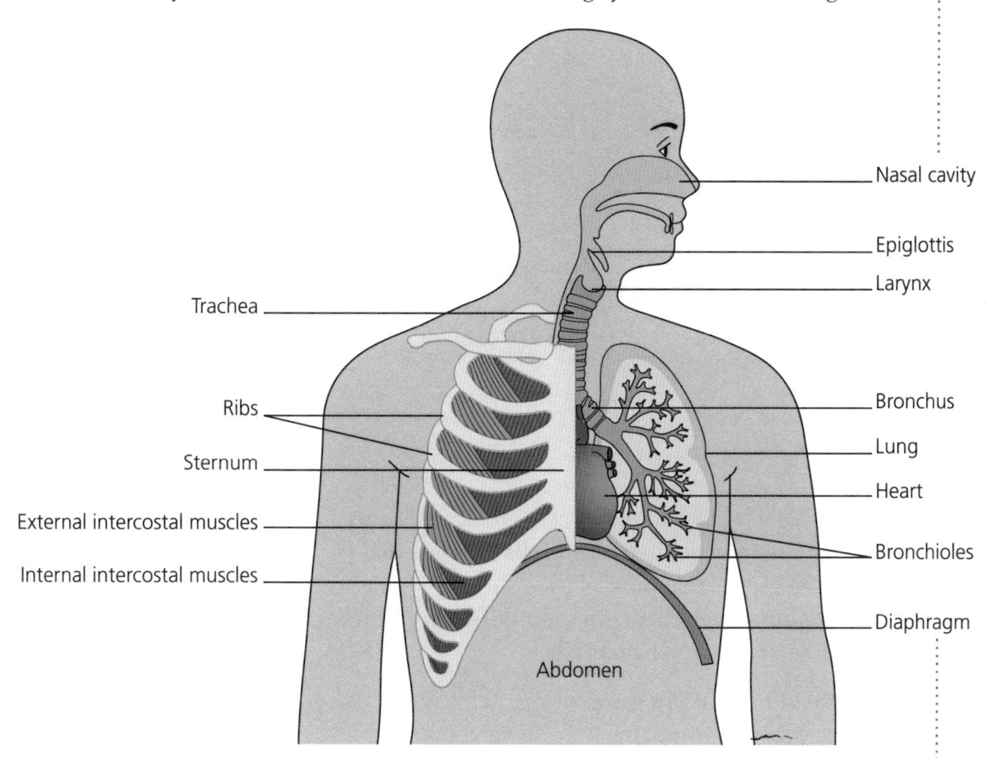

Figure 10 The human breathing system

Air is breathed in down the trachea, which is held open by rings of cartilage. The trachea branches into two bronchi, one leading to each lung. The bronchi branch into bronchioles that lead to clusters of alveoli, as shown in Figure 11.

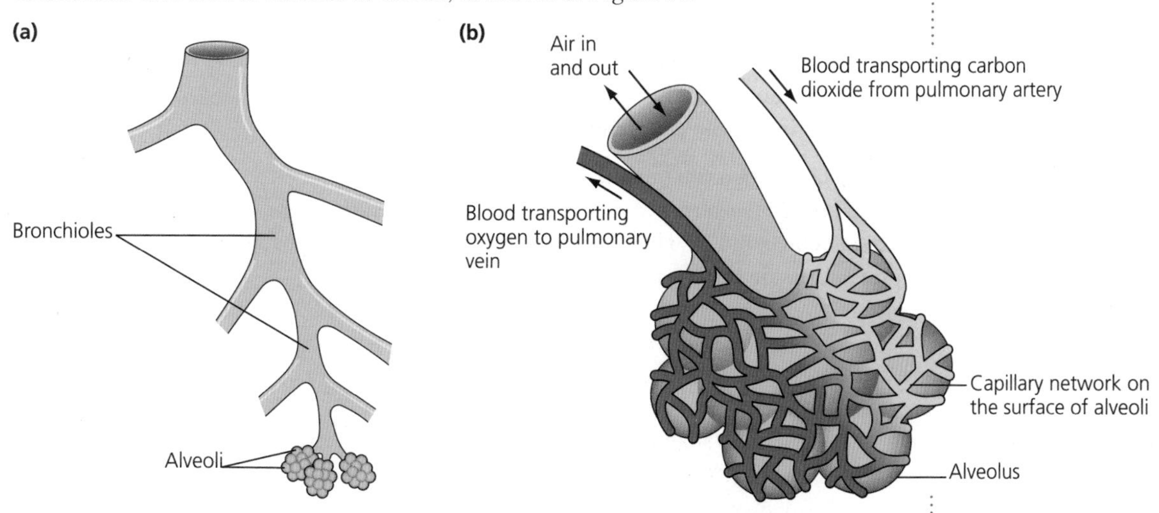

Figure 11 Bronchioles have thin branches, each ending in a cluster of alveoli surrounded by capillaries

The walls of the alveoli consist of a one-cell thick layer made of thin, flattened cells with little cytoplasm. This is shown in Figure 12.

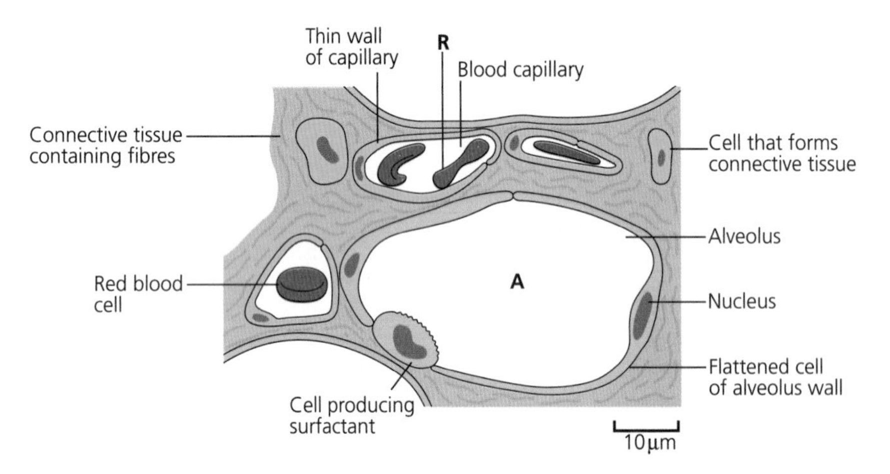

Figure 12 An alveolus and blood capillaries

Knowledge check 8

a Calculate the magnification of Figure 12.

b Calculate the actual diameter of the alveolus labelled A and the red blood cell labelled R.

The capillary walls are also made of a single layer of thin, flat cells. You can also see from Figure 12 that the capillaries are so narrow that red blood cells are squeezed through the capillaries one by one, which means that most of their surface is pressed against the capillary walls. All these features mean there is a short diffusion pathway for oxygen to pass through from the alveoli to the red blood cells.

A steep concentration gradient is required for efficient gas exchange. This is maintained by ventilation of the lungs, constantly bringing in fresh air containing a high concentration of oxygen and removing air that is high in carbon dioxide and low in oxygen. The capillaries keep removing blood that is high in oxygen and low in carbon dioxide, and bring in blood that is low in oxygen and high in carbon dioxide.

Exam tip

Don't say that the alveoli have a thin 'cell wall'. Only plants have cell walls. Instead, you should say that the walls of the alveoli are made of a single layer of thin, flattened cells.

Ventilating the lungs

Inspiration is the term for when you breathe in. When you breathe in, the **diaphragm** contracts, making it flatter. The **external intercostal muscles** between the ribs also contract, pulling the ribs up and out. This increases the volume of the chest cavity and decreases the pressure inside the chest cavity, so air enters the lungs.

Expiration is the term for breathing out. When you breathe out, the diaphragm relaxes, moving back to its original domed shape. The external intercostal muscles between the ribs also relax, allowing the ribs to fall down and in. This decreases the volume of the chest cavity and increases the pressure inside the chest cavity, so air leaves the lungs.

Exam tip

The external intercostal muscles and the diaphragm both contract at the same time, when breathing in, and relax at the same time, when breathing out. If you forget this in an exam, just put your hand on your chest and breathe in and out deeply. You will feel that the muscles contract when you breathe in. Never say that the ribs move up and down — they are bones so cannot move on their own. Be sure to mention that it is the *muscles* that bring about this movement.

If you breathe out forcefully, for example when you are exercising vigorously, there are **internal intercostal muscles** between the ribs on the inside. These contract and pull the ribs down and in, squeezing out more air from the lungs. You can see what happens during inspiration and expiration in Figure 13.

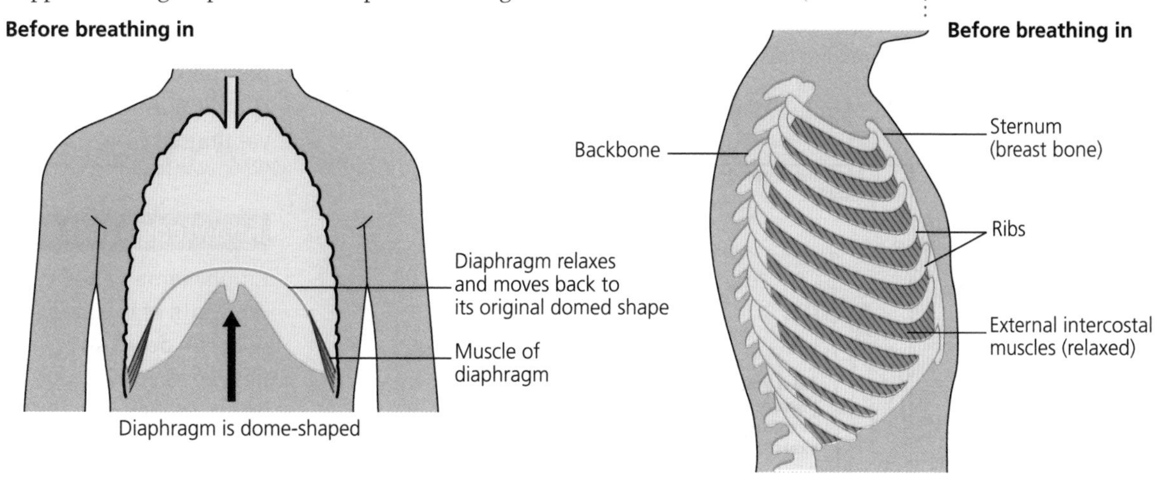

When breathing in

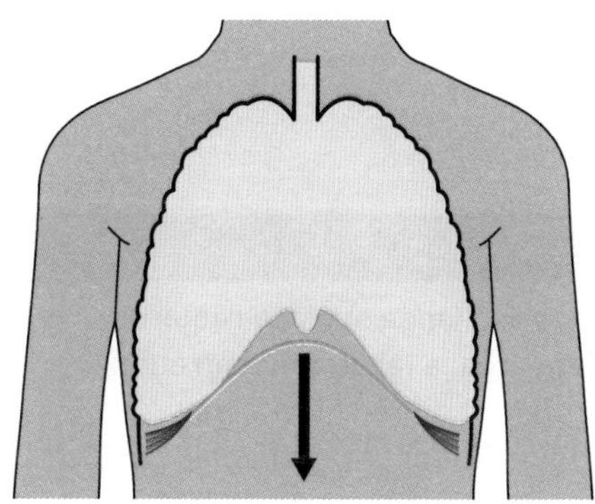

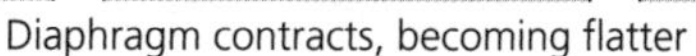

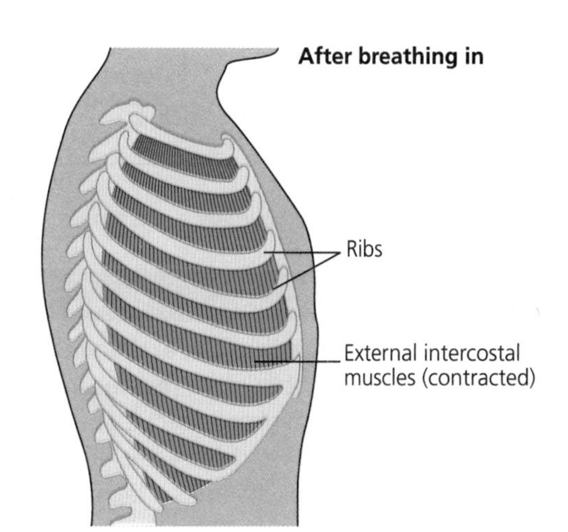

Figure 13 The position of the diaphragm (left) and the ribcage (right), before and after taking a deep breath

Exam tip

The internal and external intercostal muscles work antagonistically. The external intercostal muscles contract to pull the ribs up and out, while the internal intercostal muscles relax. The internal intercostal muscles contract to produce a forceful breath out, while the external intercostal muscles relax. Muscles work in antagonistic pairs because they can only contract (creating a 'pull') or relax — they cannot push.

We do not breathe in the same volume of air every breath. Look at Figure 14. When someone is at rest, their breathing is shallow and steady. The volume of air taken in and out in a normal breath is called the **tidal volume**. When they start to exercise, their tidal volume increases. The maximum volume someone can take into their lungs after a forced deep breath out is the **vital capacity**. There is a **residual volume** of air left in the lungs even after someone has breathed out as much as possible.

pulmonary ventilation rate = tidal volume × breathing rate

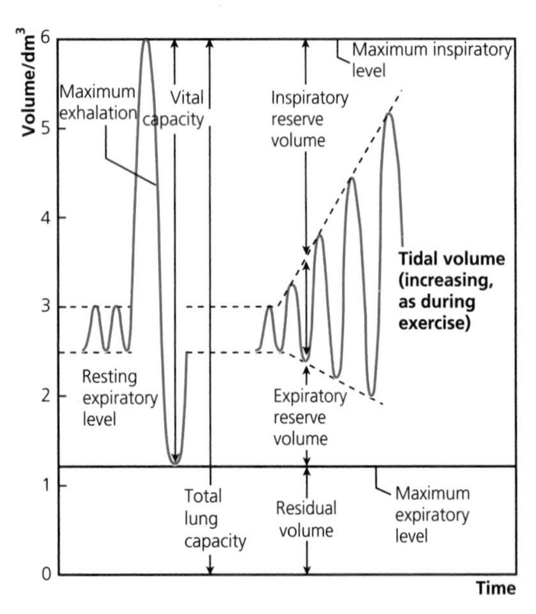

Figure 14 Lung volumes shown by a spirometer trace

Knowledge check 9

A person's tidal volume is $500\,cm^3$ and their breathing rate is 15 breaths per minute. What is their pulmonary ventilation rate?

Knowledge check 10

Emphysema causes the walls of the alveoli to break down and fibrous tissue is laid down. What effects will this have on someone's breathing and why?

Summary

- A single-celled organism can exchange gases across its whole cell-surface membrane
- The tracheal system of an insect (consisting of tracheae, tracheoles and spiracles) allows gas exchange. The gas exchange surface is where the ends of the tracheoles contact the muscle fibres.
- Gas exchange occurs in the gills of fish containing gill filaments and lamellae. Blood and water flow in opposite directions, which is known as the counter-current principle. This ensures there is a concentration gradient all the way along the exchange surface.
- The leaves of dicotyledonous plants contain mesophyll cells and stomata. Stomata allow gases to diffuse into and out of the air spaces in the leaf. The surface of the mesophyll cells is the gas exchange surface.
- An efficient gas exchange surface requires a large surface area, a large concentration gradient and a short diffusion pathway.
- Adaptations for efficient gas exchange can lead to water loss in some organisms. Insects have adaptations to reduce water loss.
- Plants adapted to live in dry conditions are called xerophytes. These have special adaptations to reduce water loss.
- The human gas exchange system consists of alveoli, bronchioles, bronchi, trachea and lungs.
- Ventilation in humans is brought about by contraction and relaxation of the intercostal muscles and diaphragm.

Digestion and absorption

Digestion takes place in the digestive system. It is when large food molecules are hydrolysed into smaller, soluble molecules that can be absorbed into the blood.

Carbohydrate digestion

Starch is digested into maltose by the enzyme **amylase**. Amylase is present in:

- saliva and secreted into the mouth
- pancreatic juice and secreted into the small intestine

Disaccharides are hydrolysed into monosaccharides by disaccharide enzymes present in the membrane of the epithelial cells of the small intestine.

- Maltase hydrolyses maltose into two molecules of glucose.
- Sucrase hydrolyses sucrose into glucose and fructose.
- Lactase hydrolyses lactose into glucose and galactose.

Glucose is absorbed from the gut lumen by co-transport as already described in Student Guide 1. Galactose is absorbed using the same transporter, whereas fructose uses a different transporter.

Lipid digestion

Lipids are hydrolysed by **lipase** enzymes secreted in pancreatic juice. This happens in the small intestine. As food passes from the stomach into the first part of the small intestine, bile from the gall bladder in the liver is secreted. This liquid contains bile salts. These bind to the fat droplets and break them down into smaller fat droplets. This is called **emulsification**. This is not digestion but a physical process that increases the surface area available for lipase enzymes to digest the lipids. Lipase secreted into the small intestine by the pancreas hydrolyses lipids into fatty acids, glycerol and monoglycerides (partly digested lipids). Fatty acids can have hydrocarbon chains of different lengths: short chain fatty acids and longer chain fatty acids.

Short chain fatty acids diffuse directly into the blood from the lumen of the small intestine via the epithelial cells. They can pass easily through the membranes because they can diffuse through the phospholipid bilayer.

Longer chain fatty acids, monoglycerides and glycerol associate with bile salts and phopholipids to form micelles. Micelles are about 200 times smaller than emulsion droplets. Micelles break down to release fatty acids and monoglycerides, which diffuse into the epithelial cells. Once inside the epithelial cells, they recombine to form triglycerides again. These triglycerides are packaged with cholesterol and phospholipids to form water-soluble fat droplets called chylomicrons. These are transferred to a lymph vessel inside the villus, called a lacteal, by exocytosis. Exocytosis is when a small piece of the cell-surface membrane is wrapped around the lipid droplets and pinched off, so that the fatty droplets are now wrapped in membrane as they enter the lymph vessels. They eventually enter the blood system.

Knowledge check 11

The amylase secreted in saliva does not remain active in the acid pH of the stomach. Use your knowledge of enzymes to explain why.

Knowledge check 12

Name the bond that is hydrolysed by maltase.

Knowledge check 13

Suggest an explanation for glucose and galactose using the same transporter, whereas fructose uses a different transporter.

Knowledge check 14

Name the bond that is hydrolysed by lipase.

Protein digestion

Proteins are digested by enzymes called **proteases**.

- In the stomach, an endopeptidase hydrolyses peptide bonds in polypeptides, breaking long polypeptides down to smaller polypeptides.
- In the small intestine, pancreatic juice secretes both an endopeptidase and an exopeptidase. The endopeptidase hydrolyses longer polypeptides into shorter polypeptides and peptides. The exopeptidase splits off dipeptides from the ends of the polypeptide and peptide chains.
- There are dipeptidase enzymes in the cell-surface membranes of the epithelial cells lining the small intestine. These hydrolyse the peptide bonds, releasing amino acids.
- The amino acids are transported into the epithelial cells by a co-transport mechanism, similar to the one that absorbs glucose. This is shown in Figure 15.

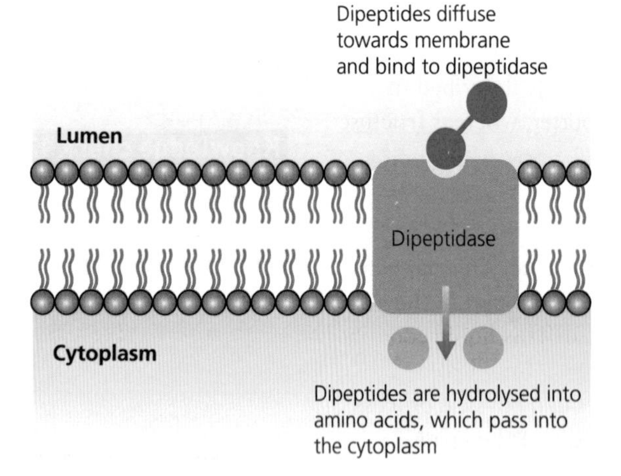

Figure 15 Dipeptidase enzymes in the cell-surface membrane of the epithelial cells of the small intestine

Knowledge check 15

In the first few days after birth, whole proteins can be absorbed into the body from the gut. However, this cannot happen after this time. Explain why this is an advantage in a newborn baby.

Summary

- During digestion, large biological molecules are hydrolysed to smaller molecules that can be absorbed across cell membranes.
- Starch is hydrolysed to maltose by amylase. Disaccharides are hydrolysed to monosaccharides by disaccharidases in the cell-surface membrane of the epithelial cells.
- Lipids are hydrolysed by lipase enzymes.
- Bile salts emulsify large fat droplets into smaller fat droplets with a larger surface area for lipase enzymes to work on.
- Amino acids and monosaccharides are absorbed into the epithelial cells by a co-transport mechanism.
- Fatty acids and monoglycerides combine with other molecules such as phospholipids to form tiny droplets called micelles. These maintain a supply of fatty acids and monoglycerides surrounding the epithelial cells.
- Fatty acids, glycerol and monoglycerides diffuse into epithelial cells where they are reassembled into triglycerides.
- The triglycerides are packaged into membrane-bound chylomicrons. These enter the lymph vessels and then the blood system.
- Proteins are hydrolysed by endopeptidases that digest polypeptides into shorter polypeptides and peptides. Exopeptidases hydrolyse peptide bonds, releasing dipeptides from the ends of peptides and polypeptides.

Mass transport

Mass transport in animals

Haemoglobin

Haemoglobin is a large protein made of four polypeptide chains. It is therefore said to have a quaternary structure. The quaternary structure of a protein is the shape of a protein that is made up of two or more polypeptide chains.

Each of the polypeptide chains has an organic non-protein group attached to it, called a haem group. Within the haem group is an Fe^{2+} ion that can combine reversibly with an oxygen molecule. You can see the structure of haemoglobin in Figure 16.

When a molecule of oxygen binds to the first haem group, the haemoglobin changes shape, making it easier for the next oxygen molecule to bind. Every time a molecule of oxygen binds to haemoglobin, the haemoglobin changes shape, making it easier for the next molecule of oxygen to bind.

Haemoglobin has a high affinity (or attraction) for oxygen. When all the haem groups have bound to an oxygen molecule, the haemoglobin molecule has four oxygen molecules attached. Haemoglobin with oxygen attached is called oxyhaemoglobin. Oxygen associates with (is attached to) haemoglobin when the concentration of oxygen is high. When the concentration of oxygen is low, oxygen dissociates (falls away) from haemoglobin.

When all the molecules of haemoglobin in the blood have four oxygen molecules attached, the haemoglobin is 100% saturated. We measure the amount of oxygen in the blood that is attached to haemoglobin as the percentage saturation. As oxygen is a gas, we measure its concentration in a mixture of gases as partial pressure. It is measured in kilopascals (kPa).

Figure 17 shows the amount of haemoglobin that is bound to oxygen at different concentrations (partial pressures) of oxygen. This is called the **oxyhaemoglobin dissociation curve**. You can see that when there is a high partial pressure of oxygen, as would happen in the lungs, almost all the haemoglobin molecules are fully saturated with oxygen. However, as the haemoglobin reaches another part of the

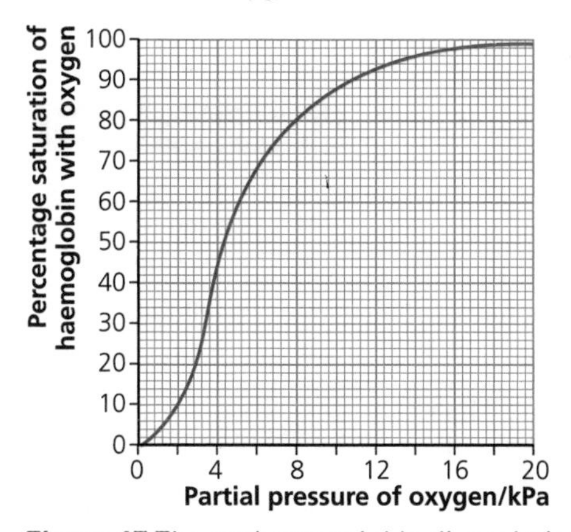

Figure 17 The oxyhaemoglobin dissociation curve for adult human haemoglobin

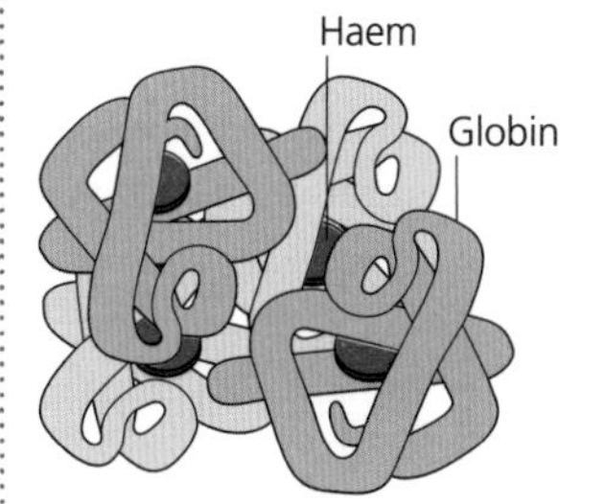

Figure 16 The structure of a haemoglobin molecule

Exam tip

Make sure you use the correct terms in your answers in exams, such as percentage saturation, partial pressure, affinity and dissociates.

Knowledge check 16

Assume the partial pressure of oxygen in the brain is 6 kPa and the partial pressure of oxygen in the liver is 4 kPa. Use Figure 17 to calculate how much more oxygen the liver receives compared to the brain.

body where there is a lower partial pressure of oxygen, the percentage saturation of the haemoglobin goes down. You can measure this off the y-axis. For example, if the partial pressure of oxygen in the lungs is 16 kPa and the partial pressure of oxygen in the brain is 6 kPa, the percentage saturation of haemoglobin falls from 98% to 70%. Therefore, the brain receives 28% of the oxygen that is carried by the haemoglobin.

Carbon dioxide decreases the oxygen-carrying ability of the blood because it reacts with water in the blood to form a weak acid called carbonic acid. This releases hydrogen (H^+) ions in the blood. Haemoglobin takes up these hydrogen ions but, as it does so, it changes shape and releases oxygen. You can see the effect of a low concentration of carbon dioxide on the oxyhaemoglobin dissociation curve in Figure 18.

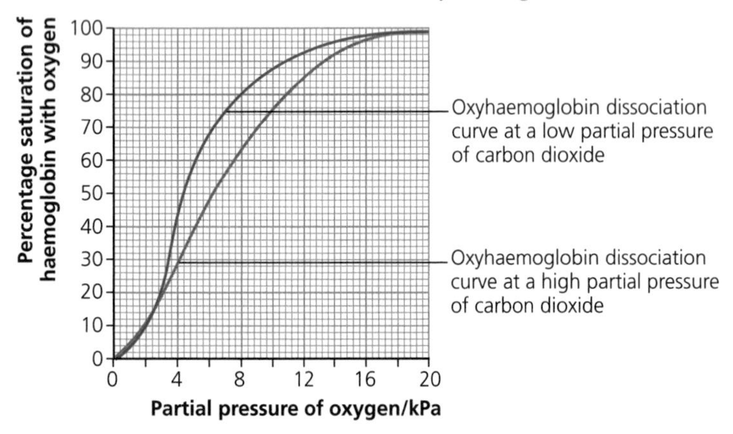

Figure 18 The oxyhaemoglobin dissociation curve at low and high partial pressures of carbon dioxide

You can see that carbon dioxide reduces the affinity of haemoglobin for oxygen. If you look at Figure 18 and read across from 50% on the y-axis, at a low partial pressure of carbon dioxide haemoglobin is 50% saturated with oxygen when the partial pressure of oxygen is 4 kPa. However, at a high partial pressure of carbon dioxide, the haemoglobin is only 50% saturated with oxygen at a partial pressure of oxygen of 6 kPa. In other words, when carbon dioxide partial pressure (concentration) is high, the haemoglobin gives up more oxygen. This is called the **Bohr effect**.

The advantage of this is that rapidly respiring tissues have a higher partial pressure of carbon dioxide, for example leg muscles in a marathon runner. The increased partial pressure of carbon dioxide causes the haemoglobin to release even more oxygen than usual, which the leg muscle cells need for respiration. Note, however, that the haemoglobin still saturates with oxygen in the lungs.

Many kinds of haemoglobin are found in different animals. They all show the same general properties just described, but they adapt different kinds of organisms to different environments. You can see the oxyhaemoglobin dissociation curves for different types of haemoglobin in Figure 19.

The type of haemoglobin that has the kind of dissociation curve represented as line A in Figure 19 is a haemoglobin with a high affinity for oxygen. You can see that the haemoglobin is fully saturated with oxygen even when the partial pressure of oxygen is low. The haemoglobin does not dissociate until the partial pressure of oxygen is low.

Exam tip

You are used to reading graphs from left to right, i.e. from the origin. Haemoglobin dissociation curves make more sense if you read them from right to left. At the right-hand side the haemoglobin is fully saturated with oxygen at the high partial pressure of oxygen found in the lungs. As the curve moves to the left, the partial pressure of oxygen drops as it meets parts of the body where there is less oxygen.

Knowledge check 17

Copy Figure 18 and add a line to show the oxyhaemoglobin dissociation curve you would expect if the partial pressure (concentration) of carbon dioxide was even higher.

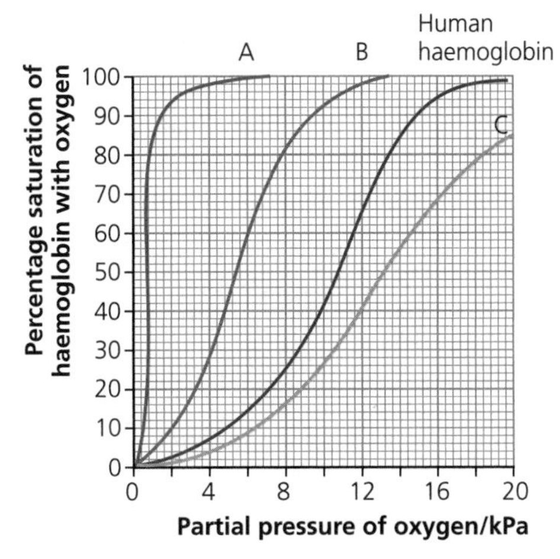

Figure 19 Oxyhaemoglobin dissociation curves for different types of haemoglobin

This kind of haemoglobin is good at acting as an **oxygen store**. An example of this is seal haemoglobin. Seals breathe in oxygen at the surface, so their haemoglobin fully saturates. However, they dive underwater where they cannot take in oxygen. Some kinds of seal stay submerged for over an hour. This type of haemoglobin means that the haemoglobin does not release its oxygen until the partial pressure of oxygen is extremely low.

The type of haemoglobin with the type of dissociation curve labelled B in Figure 19 is found in animals that live in a low oxygen environment. This type of haemoglobin saturates with oxygen at low partial pressures of oxygen and gradually dissociates as the partial pressure of oxygen falls. An example of an animal with this kind of haemoglobin is the llama. This lives at high altitude in the Andes mountains where the partial pressure of oxygen is permanently low. Another example is the human foetus. The foetus obtains oxygen from its mother's blood in the placenta. It would not receive enough oxygen if it had the same kind of haemoglobin as its mother. You can see that the foetus's haemoglobin can become fully saturated with oxygen in the placenta even if the mother's blood is not fully saturated, because this type of haemoglobin has a higher affinity for oxygen than adult haemoglobin.

The type of haemoglobin labelled C has a lower affinity for oxygen than adult human haemoglobin. This kind of haemoglobin dissociates readily. It is typical of organisms that need a great deal of oxygen because they have a high metabolic rate, but they live in an environment where oxygen is plentiful. An example is the haemoglobin of small mammals such as mice or shrews. These are small animals that lose a lot of body heat, so they have a high rate of respiration. However, they live in an oxygen-rich environment so they do not need a haemoglobin that stores oxygen.

The circulation of the blood

Figure 20 shows the general pattern of blood circulation in a mammal such as a human.

You will see that it is a **double circulation**. This means that the deoxygenated blood returning to the heart is pumped to the lungs to be oxygenated, but it then returns to the heart to be pumped all around the body.

Knowledge check 18

If the line labelled B in Figure 19 is for llama haemoglobin, what shape of oxyhaemoglobin dissociation curve would you expect for a llama foetus?

Knowledge check 19

Carbon monoxide, produced by badly maintained gas appliances or by cigarette smoking, binds to the haem groups in haemoglobin much more readily than oxygen. Use this to explain how carbon monoxide can cause death by asphyxia and why pregnant women are advised not to smoke.

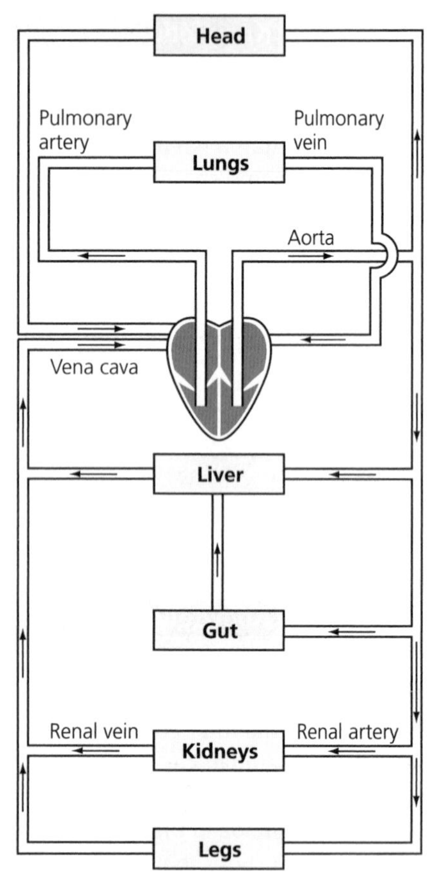

Figure 20 The general pattern of blood circulation in a mammal

> **Exam tip**
> You only need to know the names of the blood vessels labelled in Figure 20 and the coronary arteries that supply the heart muscle with glucose and oxygen (although these are not shown in the diagram).

The heart

Figure 21 shows the external view of the heart. The walls of the heart are made of **cardiac muscle**. You can see that there are **coronary arteries** supplying oxygen and glucose to the cardiac muscle.

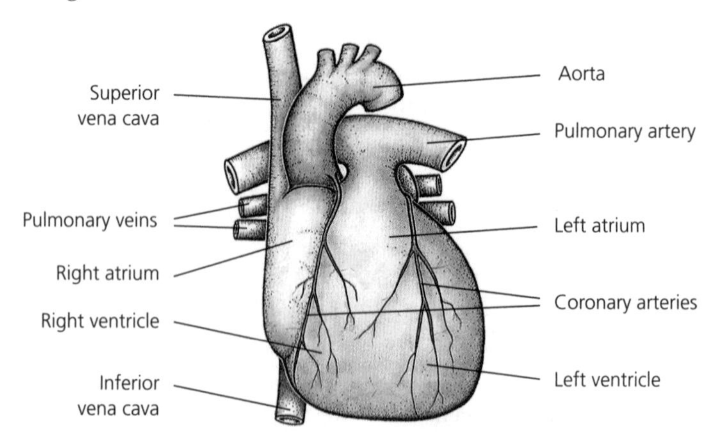

Figure 21 The external view of the heart

Figure 22 shows a vertical section through the heart. It is made up of four chambers. The right atrium and right ventricle contain deoxygenated blood that has been round the body, and is returned to the right atrium via the vena cava. The right ventricle pumps this blood to the lungs to be oxygenated via the pulmonary artery. The oxygenated blood returns to the heart from the lungs to the left atrium of the heart via the pulmonary vein. The left ventricle then pumps this oxygenated blood round the body via the aorta.

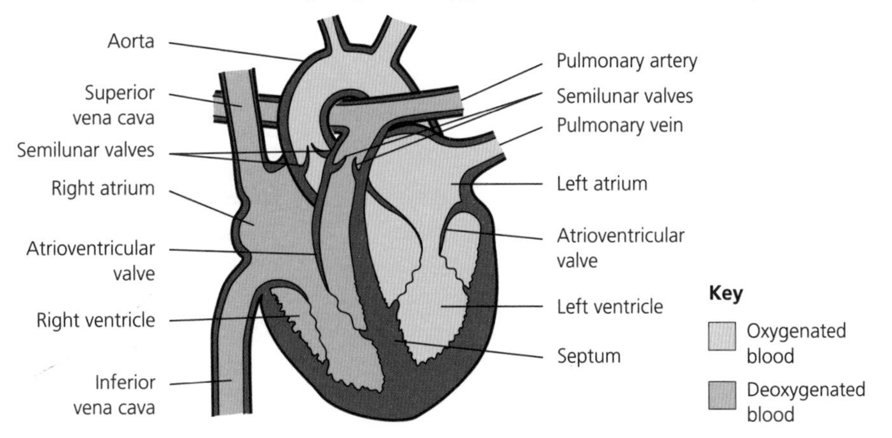

Figure 22 A vertical section through the heart

Knowledge check 20

In Figure 22 you can see that there are string-like tendons attached to the atrioventricular valves. Suggest why these are important.

When the cardiac muscle contracts, the walls of the heart squeeze in on the blood in the heart and pump the blood. There are valves to ensure that the blood flows in one direction only. The left and right sides of the heart contract and relax at the same time as each other. This pumping action is called the **cardiac cycle** and is shown in Figure 23.

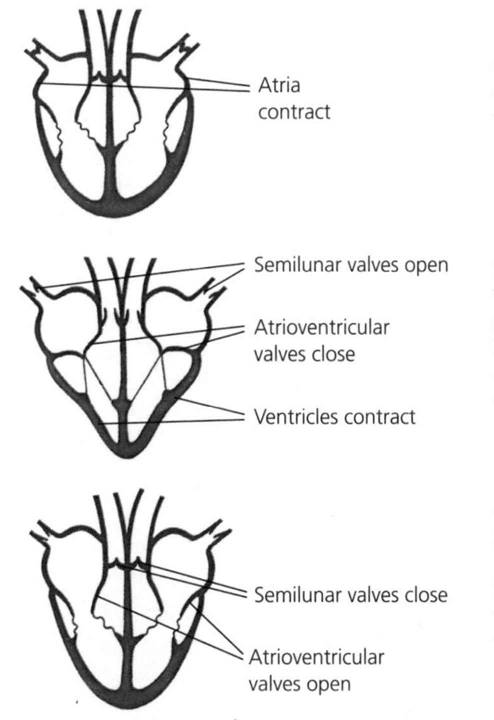

During **atrial systole**, the muscle in the walls of the atria contracts, pushing more blood into the ventricles.

During **ventricular systole**, the muscle in the walls of the ventricles contracts. This causes the pressure of the blood inside the ventricles to become greater than in the atria, forcing the atrioventricular valves shut. The blood is forced out through the aorta and pulmonary artery.

During **diastole**, the heart musles relax. The pressure inside the ventricles becomes less than inside the aorta and pulmonary artery, so the blood inside these vessels pushes the semilunar valves shut. Blood flows into the atria from the veins, so the cycle is ready to begin again.

Figure 23 The stages of the cardiac cycle

Exam tip

Don't confuse the pulmonary artery and the pulmonary vein. **A**rteries always carry blood **a**way from the heart, and vei**ns** carry blood back **in**to the heart. However, the pulmonary artery is the only artery that carries deoxygenated blood, as it takes blood away from the right ventricle to the lungs. The pulmonary vein is the only vein that carries oxygenated blood, as it brings oxygenated blood from the lungs back into the left atrium.

The contraction of the cardiac muscle that makes up the walls of the heart chambers creates pressure inside the heart, but when the muscle relaxes the pressure falls. Figure 24 shows how the pressure changes in the aorta, left ventricle, right ventricle and left atrium during the cardiac cycle.

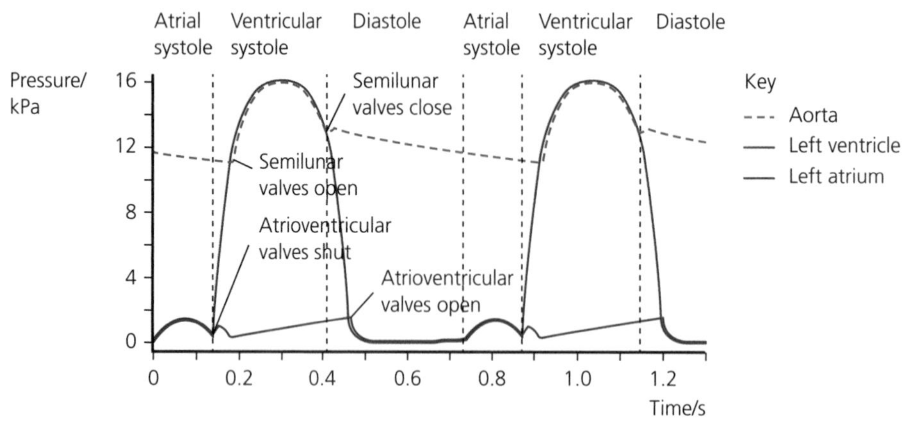

Figure 24 Pressure changes in the heart during the cardiac cycle

> **Knowledge check 21**
>
> What is the heart rate of the person whose cardiac cycle is shown in Figure 24?

> **Knowledge check 22**
>
> If a curve showing the pressure in the right ventricle was added to Figure 24, what shape would it have?

Remember that:

- The atrioventricular valves open when the pressure in the atrium is greater than the pressure in the ventricle.
- The atrioventricular valves close when the pressure in the ventricle is greater than the pressure in the atrium.
- The semilunar valves open when the pressure in the ventricle is greater than the pressure in the aorta.
- The semilunar valves close when the pressure in the aorta is greater than the pressure in the ventricle.

> **Exam tip**
>
> Remember that artery walls are elastic and that they show elastic recoil to keep the blood under pressure. They do not contract and relax, nor do they push the blood along.

Blood vessels

Arteries carry blood away from the heart. The blood passing along them is at high pressure and pulsing because it is being pumped by the heart. To keep the blood under pressure, the arteries have thick walls with elastic tissue, so that the wall can stretch and recoil as the blood flows through. There are also muscle fibres in the walls. Arteries branch into smaller **arterioles**. These also contain smooth muscle in their walls, which can contract or relax to regulate the flow of blood through them.

Veins carry blood back to the heart. Blood in the veins is at low pressure so the walls are thinner than those of arteries, with less muscle and elastic tissue. The lumen is larger than that of arteries. Veins contain valves to make sure that there is no backflow of blood.

> **Exam tip**
>
> Don't say that capillaries have thin cell walls — animal cells don't have cell walls. You need to say that the wall of the capillary is thin because it is made of a single layer of thin, flattened cells.

Capillaries are tiny blood vessels that are extremely narrow. They are just wide enough for red blood cells to squeeze through. Their walls are made of a single layer of thin, flattened endothelial cells. There are often gaps between the cells that allow plasma (the liquid part of the blood) to leak out. Capillaries have thin walls and a narrow lumen so that they can exchange substances rapidly between the blood and the body cells, such as oxygen and nutrients.

The structure of these blood vessels is shown in Figure 25. Note that the capillary is not drawn to the same scale as the artery and vein.

Pressure changes in the circulatory system

Figure 26 shows the changes in the pressure of the blood as it flows through the circulatory system.

- Blood in the arteries is under high pressure because it has just been pumped out of the left ventricle of the heart. The pressure goes up and down in time with systole and diastole of the cardiac cycle. The elastic fibres in the artery walls stretch and recoil, keeping the blood under pressure and smoothing out the surges in pressure. The pressure gradually drops as blood flows through the arteries and arterioles because of friction between the blood and the artery walls.
- The pressure drops in the capillaries because their total cross-sectional area is greater than the total cross-sectional area of the arteries.
- The blood is at its lowest pressure in the veins because it is now a long way from the pumping action of the heart. Having a wide lumen reduces the effects of friction which would otherwise slow the blood down further.

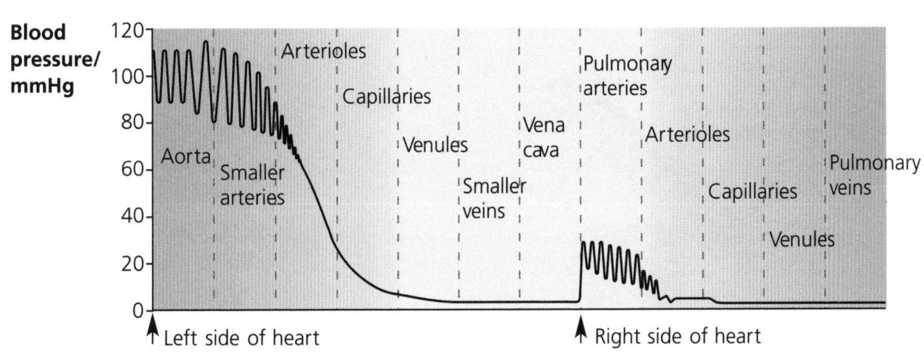

Figure 26 Pressure changes in the blood as it flows through the circulatory system

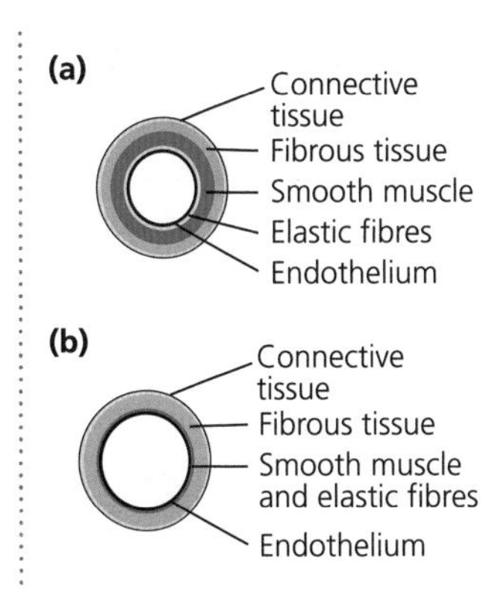

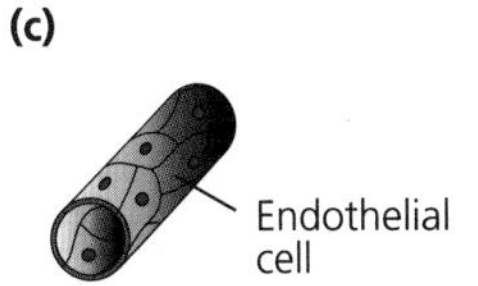

Figure 25 The structure of (a) arteries, (b) veins and (c) capillaries

Knowledge check 23

The graph shows the rate of blood flow and the total cross-sectional area of different blood vessels. Use the information in the graph to explain why the rate of blood flow is higher in veins than in capillaries.

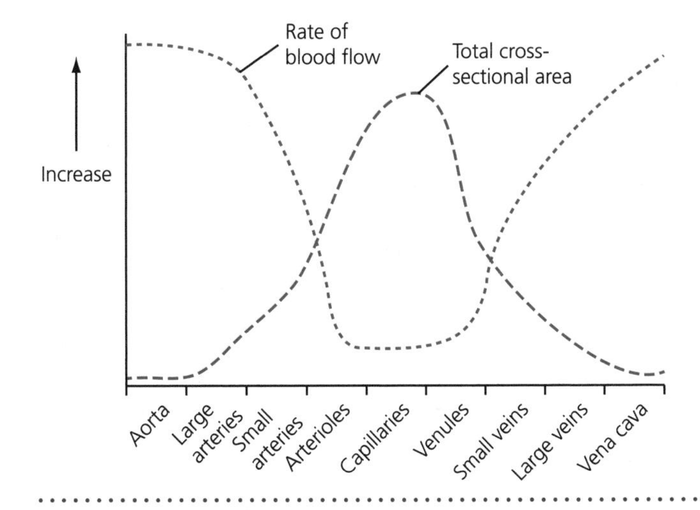

Tissue fluid formation

Tissue fluid is an intercellular fluid that forms when water and small molecules leave the blood capillaries. It transports oxygen and nutrients to body cells, and removes carbon dioxide and other waste products. You can see this process in Figure 27.

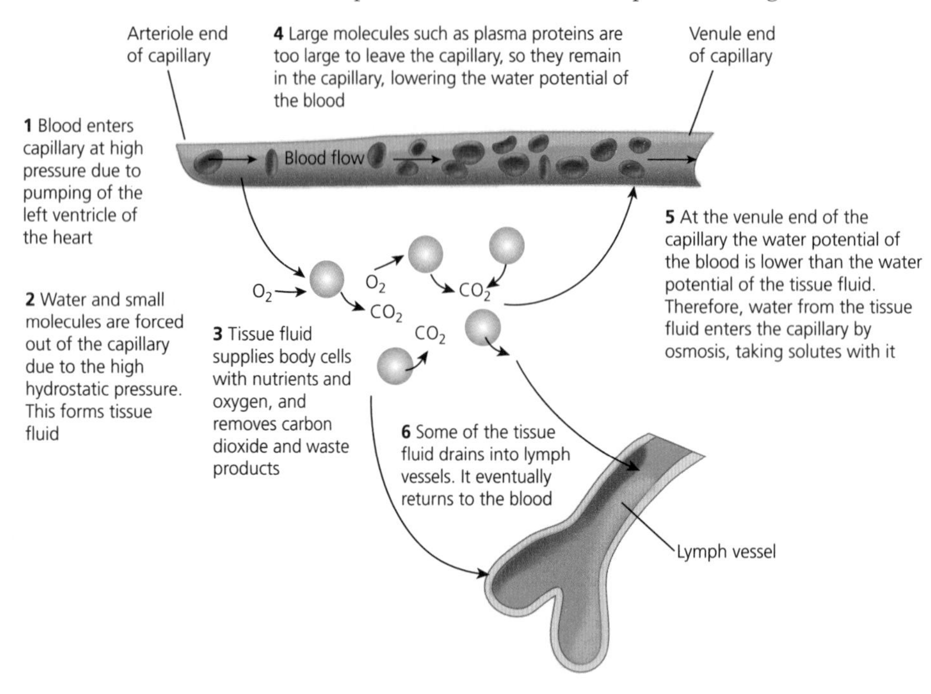

Figure 27 How tissue fluid is formed and returned to the blood

Blood arrives at a capillary network from an arteriole. The blood is at high pressure from the pumping action of the heart. This creates a **hydrostatic pressure**. Molecules small enough to pass through the pores in the wall of the capillary are squeezed out in a process called **ultrafiltration**.

This means that water from the blood plasma, together with small molecules such as dissolved oxygen and glucose, pass out of the capillary to form **tissue fluid**. However, larger molecules such as the plasma proteins remain in the capillary.

At the venule end of the capillary, the water potential of the blood is lower than the water potential of the tissue fluid. Therefore, the water from the tissue fluid diffuses back into the capillary at the venule end by osmosis, carrying solutes with it. However, not all the tissue fluid can be reabsorbed this way. The rest of the tissue fluid drains into blind-ended lymph vessels. These will eventually return the fluid in them — called lymph — to the blood system. If the lymph vessels become blocked, fluid builds up in the tissues. This causes swelling called **oedema**.

Exam tip

Remember, *only water* can move by osmosis. Be careful you don't say that the tissue fluid is reabsorbed by osmosis — it is the water from the tissue fluid that is absorbed back into the capillary by osmosis.

Exam tip

'Ultrafiltration' means filtering under pressure. It is rather like what happens when you make coffee using a cafetière. The coffee passes through the holes in the filter when you press down on it, but the coffee grounds are too big to pass through.

Knowledge check 24

People who have high blood pressure often have oedema, especially in their arms and legs. Explain why.

Knowledge check 25

Kwashiorkor is a condition caused by a protein deficiency in the diet. Children with kwashiorkor have oedema that makes their abdomen swell. Explain how this happens.

Required practical 5

Dissection of animal or plant gas exchange system or mass transport system or of organ within such a system

Heart dissection

You should refer to the textbook for photographs of a heart dissection or any work you have done with your teacher.

If you wish to carry out a dissection yourself, there are some useful dissection guides on the internet, most of which show step-by-step instructions. Always obtain your material for dissection from a reputable butcher or supermarket and dispose of it carefully afterwards. Wash your hands thoroughly before and after the dissection, and wear disposable gloves. Remember that dissection instruments are sharp, so take care when using them. When using a scalpel, always cut away from your hand so that you don't cut yourself. Always place dissection instruments in disinfectant after use before you, or a technician, cleans and dries them.

A helpful guide to dissecting lungs (written for teachers but useful for students) may be found at www.nuffieldfoundation.org/practical-biology/dissecting-lungs and a similar guide for dissecting a heart is found at www.nuffieldfoundation.org/practical-biology/looking-heart

Summary

- The haemoglobins are a group of chemically similar molecules found in many organisms. Haemoglobin is a protein with a quaternary structure. It consists of four polypeptide chains.
- Haemoglobin is found inside red blood cells and is used in the transport of oxygen.
- Haemoglobin loads with oxygen when the partial pressure of oxygen is high. It transports oxygen round the body and unloads oxygen when the partial pressure of oxygen is low. This can be shown on an oxyhaemoglobin dissociation curve.
- When molecules of oxygen bind to the haem group in one of the polypeptide chains of haemoglobin, this changes the shape of haemoglobin slightly, making it easier for the next molecule of oxygen to bind.
- Carbon dioxide changes the oxyhaemoglobin dissociation curve so it shifts to the right. This means that oxyhaemoglobin dissociates even more readily. This is called the Bohr effect.
- Many animals are adapted to their environment by possessing different types of haemoglobin with different oxygen transport properties.
- Mammals have a double circulation. This means that when blood has been pumped to the lungs to become oxygenated, it returns to the heart to be pumped again around the systemic circulation.
- The heart contains two atria and two ventricles. The left side of the heart pumps oxygenated blood while the right side of the heart pumps deoxygenated blood.
- The cardiac cycle consists of atrial systole, ventricular systole and diastole. Using these stages, there are pressure and volume changes in the heart with associated valve movements that cause the blood to flow in one direction only.
- Arteries, veins and capillaries have structures that are related to their functions.
- Tissue fluid is formed in capillary beds from blood plasma. Tissue fluid supplies the cells with oxygen and nutrients. Tissue fluid is then returned to the blood via the capillaries or the lymph vessels.

Mass transport in plants

The cohesion–tension theory

Xylem vessels are long tubes that transport water from the roots of a plant to all the other parts of the plant. Their cell wall is impregnated with **lignin**, which is waterproof and strong. They have no living contents in the cell, so there is no resistance to water flow through them. The end walls of the xylem vessels break down, so they form a continuous tube. There are pores in their side walls called **pits** that allow water to move sideways out of the xylem. You can see the structure of xylem vessels in Figure 28.

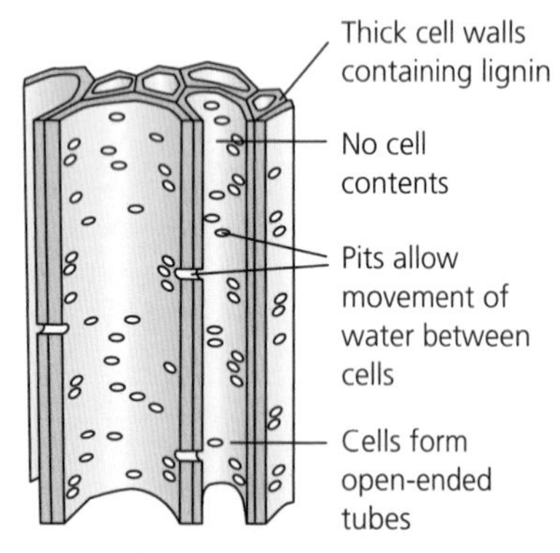

Figure 28 A longitudinal section of xylem

Getting water in

Water enters a plant through the roots. A thin film of water surrounds the particles in the soil. This water has a higher water potential than the root hair cell, so water enters the root hair cell by osmosis down a water potential gradient. The water potential of the root hair cell is now higher than the water potential of the cortex cell next to it. Therefore, water enters this cell by osmosis down a water potential gradient. In this way water can pass from the water in the soil across the root. You can see this in Figure 29.

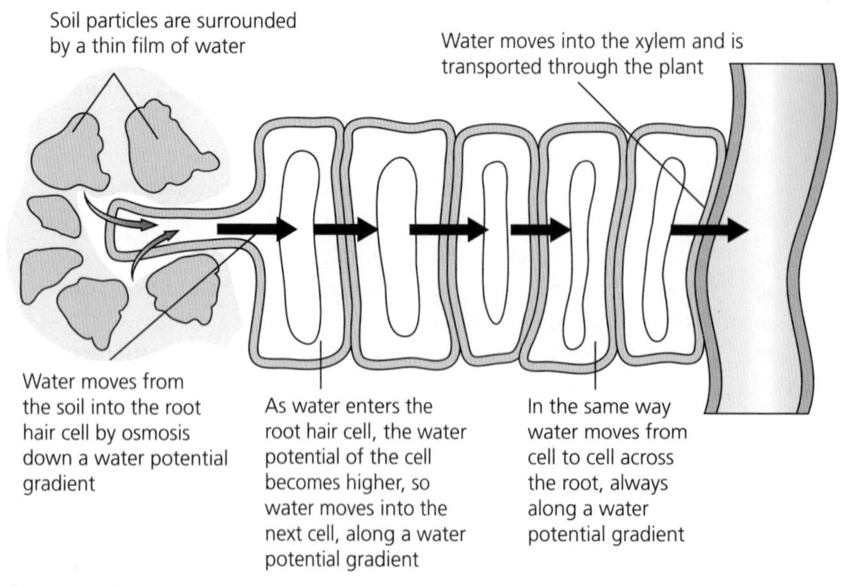

Figure 29 The path of water from the soil to the xylem

In fact, water can move across the root in two ways, as shown in Figure 30. In the **symplast pathway**, water enters the root hair cell by osmosis. The water passes from the cytoplasm of one cell to another by **plasmodesmata** (*singular*: plasmodesma), which are strands of cytoplasm linking one cell to another that pass through pores in the cellulose cell walls. There is a water potential gradient all the way across the root because the xylem is constantly moving water up the stem.

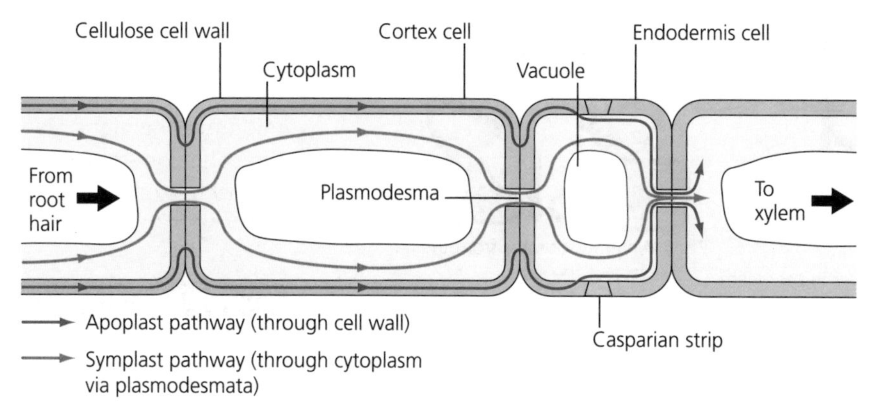

Figure 30 The symplast and apoplast pathways by which water moves through the cells of the cortex and endodermis

The cellulose cell wall is fairly rigid but not solid. It is made of a mesh of cellulose fibres so there are spaces between the cellulose fibres. In the **apoplast pathway**, water entering the root enters these spaces in the cell walls and the intercellular spaces. Hydrogen bonding attracts water molecules to each other, so as water is moved up the xylem, this draws water across the root. Water can move by this apoplast pathway until it meets the **endodermis**. This is a layer of specialised cells that surround the xylem. Endodermis cells have a waxy substance called suberin, which impregnates most of its cell wall. This is called a **Casparian strip** and it stops water moving through the apoplast pathway. Therefore, all the water now travels via the plasmodesmata following the symplast pathway. You can see this in Figure 31.

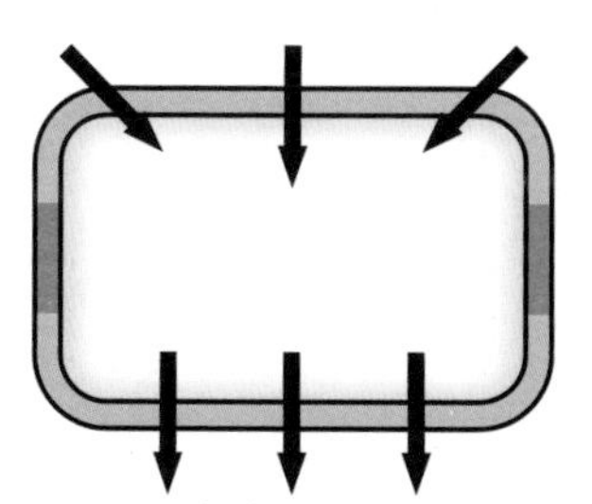

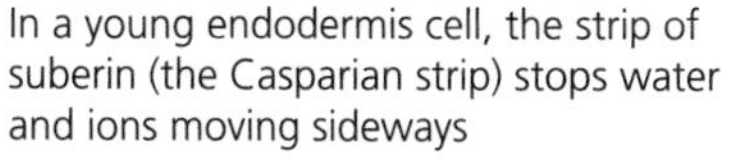

In a young endodermis cell, the strip of suberin (the Casparian strip) stops water and ions moving sideways

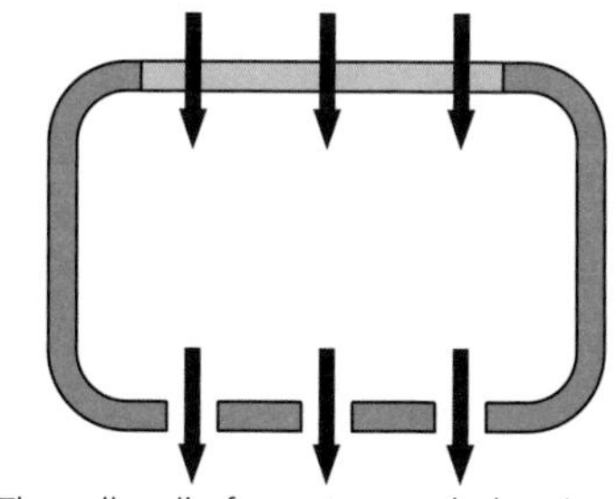

The cell wall of a mature endodermis cell is heavily suberised, so water and ions can only move onwards through the plasmodesmata

Figure 31 The role of the endodermis

Pulling water up

There are air spaces inside the leaf of a plant. These are always saturated with water vapour, which has evaporated from the surface of the mesophyll cells. This evaporation is called transpiration. As this water evaporates, it causes more water to be drawn into the cell by osmosis from its neighbouring cell, and so on. This creates a 'pull' (or tension) on the column of water in the xylem itself. The molecules of water show cohesion because of their hydrogen bonding. This makes the column of water in the xylem extremely strong and hard to break. As transpiration causes water to move out of the xylem in the leaves, this creates a tension that pulls the whole column of water in the xylem upwards. Therefore, this explanation for the movement of water through a plant is called the cohesion–tension theory and it is summarised in Figure 32.

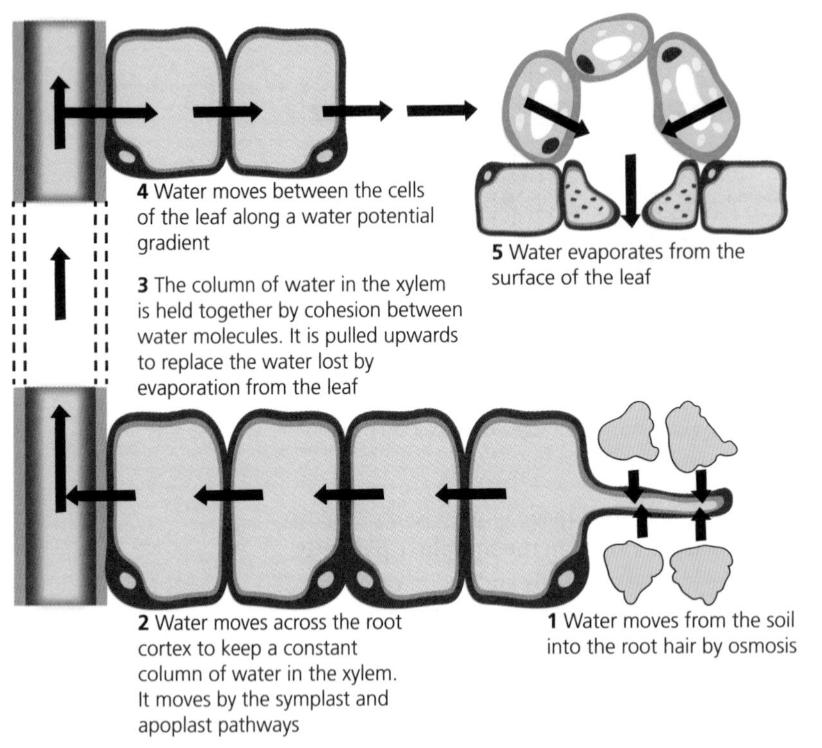

Figure 32 The cohesion–tension theory

Measuring transpiration

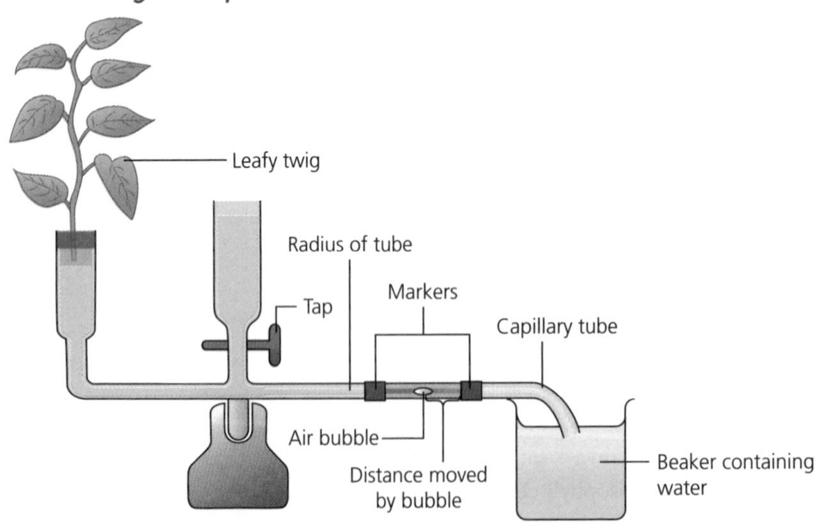

Figure 33 A potometer

Knowledge check 26

A scientist measured the diameter of a tree trunk every 30 minutes for 24 hours. The diameter was 0.2 mm less during the day, which was hot and sunny, than it was in the middle of the night. Use the cohesion–tension theory to explain why.

Knowledge check 27

When you buy a bunch of cut flowers, it is recommended that you cut off about 3 cm from the bottom of the stems, making sure they are underwater when you cut. Use the cohesion–tension theory to explain the reason for this advice.

One way to measure water loss from a plant is to use a potometer, as shown in Figure 33. Potometers measure water uptake rather than transpiration and the shoots are given an unlimited supply of water. However, they are useful for comparing water loss in different conditions or for comparing water loss from different types of plant.

Knowledge check 28

How could you find the volume of water evaporated from a shoot per minute, using the apparatus shown in Figure 33?

Mass flow in the phloem

Phloem is made up of two kinds of cell. **Sieve tubes** are long, thin cells that join together to form long tubes. At the end of the cells there are **sieve plates** with large pores in them. When sieve tubes are mature, they have few cell contents. Alongside the sieve tubes are **companion cells**. These are very active cells. The companion cells load and unload organic solutes, such as sucrose and amino acids, into the sieve tubes, while the sieve tubes transport them through the plant. You can see the structure of phloem cells in Figure 34.

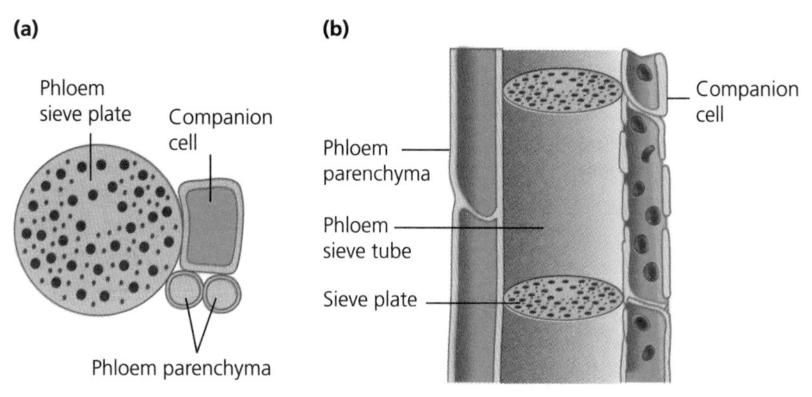

Figure 34 The structure of phloem

Cells that photosynthesise and make sugars are called **sources**. These sugars are transported to parts of the plant where they are used or stored, known as **sinks**. The sugars produced in the sources are converted into sucrose, which is transported in the phloem sieve tubes.

One theory to explain how sucrose is transported in the phloem is the **mass flow hypothesis**. According to this theory, sugars build up in the source, lowering the water potential of the source cells. This causes water to enter the cells by osmosis. These sugars are then actively transported into the phloem sieve tubes by specially adapted phloem parenchyma cells called **transfer cells**. This lowers the water potential in the phloem sieve tube, so water enters by osmosis. This causes a high **pressure potential** at the source. At the sink, sugars are removed from the phloem sieve tube by the companion cells. This raises the water potential in the sieve tube and lowers the pressure potential. Therefore, sugars in solution move by **mass flow** from the source to the sink. The mass flow theory is summarised in Figure 35.

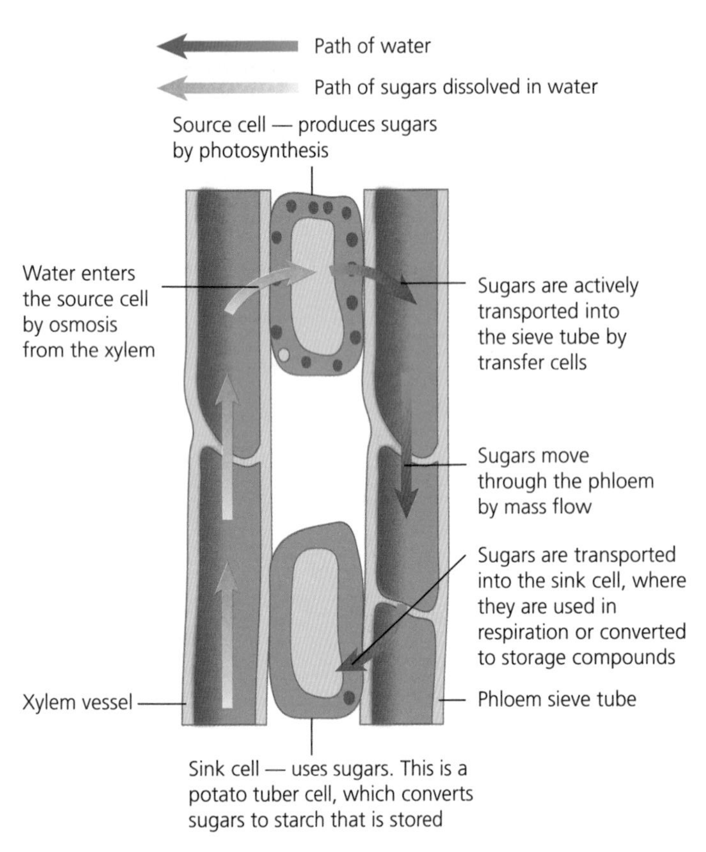

Figure 35 Mass flow hypothesis

Knowledge check 29

State whether the following are sources or sinks:

a flower buds
b spongy mesophyll cells
c onion bulbs

Investigating transport in plants

Sometimes, the phloem is sampled using aphids. These are tiny insects with a feeding tube called a proboscis, which they insert into the phloem to feed on its sugary contents. Scientists can anaesthetise the aphid and cut across its mouthparts, leaving a tiny sampling tube so that they can collect the contents of the phloem and study them. You can see this in Figure 36.

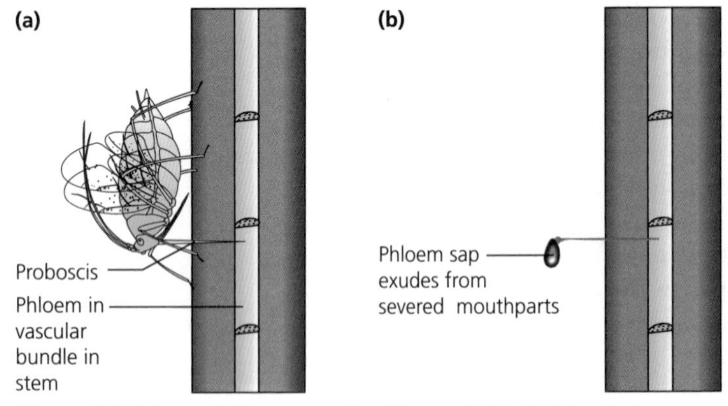

Figure 36 Using aphids to sample phloem. (a) Aphid feeds by inserting its proboscis into the phloem. (b) Aphid is anaesthetised and its head cut off from its mouthparts

Radioactive tracers can be used to investigate transport in plants. The plant makes organic substances during photosynthesis, which travel in the phloem. If the plant is supplied with carbon dioxide containing radioactive carbon (^{14}C), this becomes incorporated into organic substances in the plant. It is then possible to trace what happens to the organic substances made in photosynthesis. You can see how radioactive tracers are used in Figure 37.

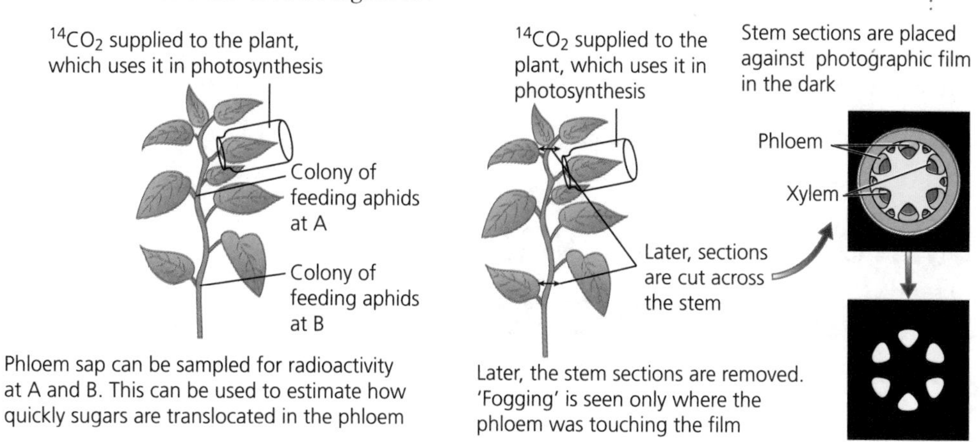

Figure 37 Using radioactive tracers

Another technique used to investigate transport in plants is ringing. Cylinders of bark and underlying tissue are removed from plant stems. This removes the phloem and outer cells from the stem, but leaves the xylem intact. The contents of the phloem above and below the ring can be sampled using a technique such as aphid mouthparts. You can see this in Figure 38.

Knowledge check 30

Why is it important to leave the xylem intact in a ringing investigation?

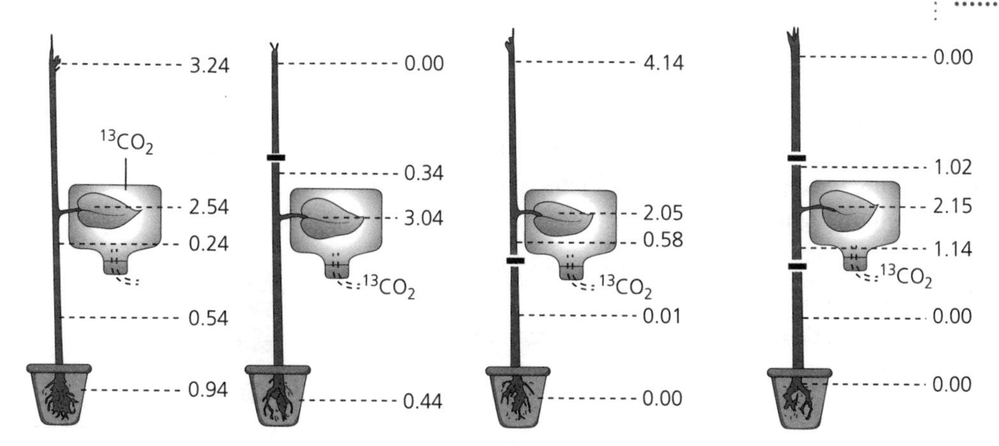

Figure 38 The results of a ringing experiment using radioactive carbon dioxide

Knowledge check 31

How do the results in Figure 38 support the mass flow hypothesis?

Exam tip

You don't need to know any specific investigation, but you do need to understand the use of radioactive tracers and ringing experiments. You need to be able to evaluate investigations and their results, and be able to explain whether or not they support the standard hypotheses.

Summary

- The cohesion–tension theory explains how water moves through a plant. Water enters the root hair cells by osmosis and then travels across the root cortex by the symplast or apoplast pathways.
- Evaporation of water from the leaves (transpiration) creates a 'pull' that causes a column of water to flow upwards through the xylem tissue.
- Sucrose and other organic solutes travel from sources (where they are made) to sinks (where they are used up or stored) in phloem tissue.
- The mass flow hypothesis explains how substances travel in the phloem.
- Transport in plants can be investigated using radioactive tracers and ringing experiments.

Genetic information, variation and relationships between organisms

DNA, genes and chromosomes

DNA in prokaryotes and eukaryotes

In prokaryotes, DNA is short, circular and not associated with proteins. In eukaryotes, DNA is long and linear, and wound around proteins called histones to form **chromosomes**.

Chloroplasts and **mitochondria**, both organelles found in eukaryotic cells, also contain their own DNA. This DNA is like the DNA in prokaryotic cells, i.e. it is short, circular and not complexed with proteins.

Knowledge check 32

Name the small, circular pieces of DNA found in some prokaryotic cells that carry genes additional to those in the main piece of circular DNA.

A gene is a base sequence of DNA that codes for the amino acid sequence of a polypeptide or for a functional molecule of RNA, whether mRNA, tRNA or ribosomal RNA. A gene is found in a specific fixed position, called a **locus** (*plural:* loci), on a particular chromosome, as shown in Figure 39.

A specific sequence of three bases in DNA, called a triplet, codes for a specific amino acid. The genetic code has three important properties. It is:

- Universal — the DNA code is exactly the same in all living organisms.
- Non-overlapping — the DNA code is 'read' three bases at a time, then the next three. In other words, each base is part of only one triplet.
- Degenerate — there are more triplets than are necessary to code for 20 amino acids. Therefore, some amino acids are coded for by several different triplets, whereas other amino acids are coded for by one triplet only.

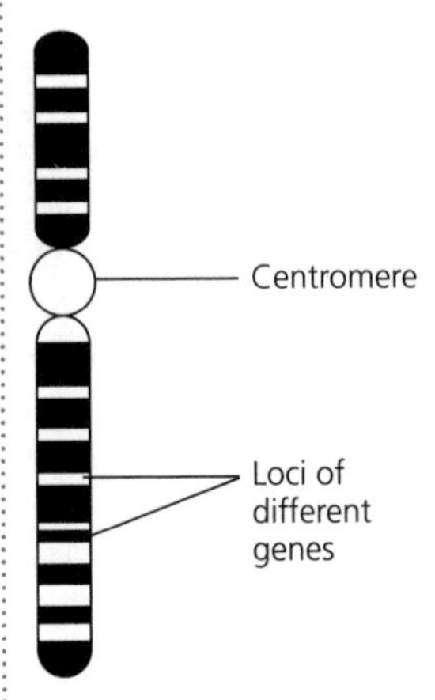

Figure 39 A chromosome

We now know that not all the DNA in the chromosomes of eukaryotes codes for polypeptides (see Figure 40):

- Between the genes, there are non-coding multiple repeats of base sequences.
- Within the genes, there are sequences (exons) that code for sequences of amino acids. There are also introns, which are non-coding regions.

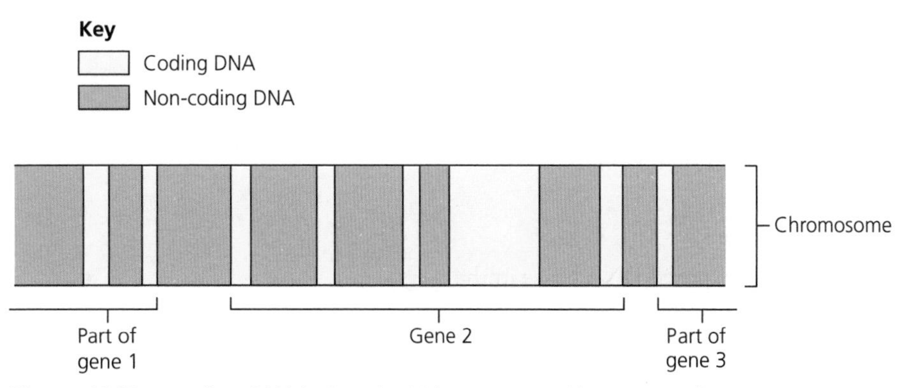

Figure 40 Non-coding DNA is found within a gene and between adjacent genes

DNA and protein synthesis

The **genome** is a term for the complete set of genes in a cell. The **proteome** is the full range of proteins that a cell is able to produce.

DNA carries the code for proteins, but synthesising proteins also involves RNA. Two main kinds of RNA are involved:

- Messenger RNA (mRNA) is a single-stranded polynucleotide consisting of hundreds or thousands of RNA nucleotides, complementary to the base sequence of the DNA in the gene.
- Transfer RNA (tRNA) is a single-stranded polynucleotide, about 75 nucleotides long. It coils up into a clover-leaf shape by forming hydrogen bonds between some of the bases. It carries specific amino acids to the ribosome. At one part of the molecule there are three exposed bases called the anticodon, which are complementary to one of the mRNA codons. At the other end of the molecule there are three exposed bases, which form an amino acid attachment site. This carries a specific amino acid determined by the anticodon.

These two kinds of RNA are shown in Figure 41. Ribosomal DNA (rDNA), which is synthesised by genes in the nucleolus, forms part of the structure of the ribosomes.

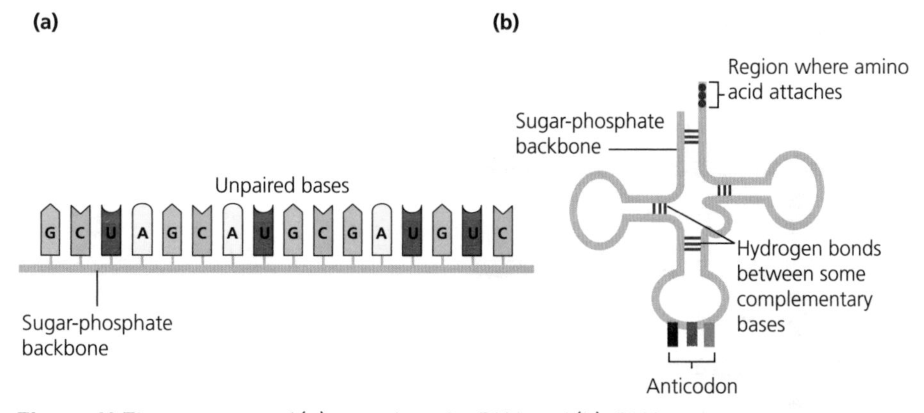

Figure 41 The structure of (a) a section of mRNA and (b) tRNA

Exam tip

Remember that tRNA is not double-stranded. It consists of a single strand of RNA that folds up and forms hydrogen bonds between bases in different parts of the chain.

Protein synthesis

Transcription

Transcription is the first stage of protein synthesis and it occurs in the cell nucleus:

1 The section of DNA to be transcribed (the gene) unwinds and the hydrogen bonds between the bases break.
2 One strand of the DNA becomes a template.
3 New RNA nucleotides align alongside the exposed DNA bases in the template strand, by complementary base-pairing.
4 The enzyme RNA polymerase joins the RNA nucleotides together to form a molecule of mRNA complementary to the DNA.
5 The completed mRNA leaves the nucleus via a pore in the nuclear membrane.

Transcription is shown in Figure 42.

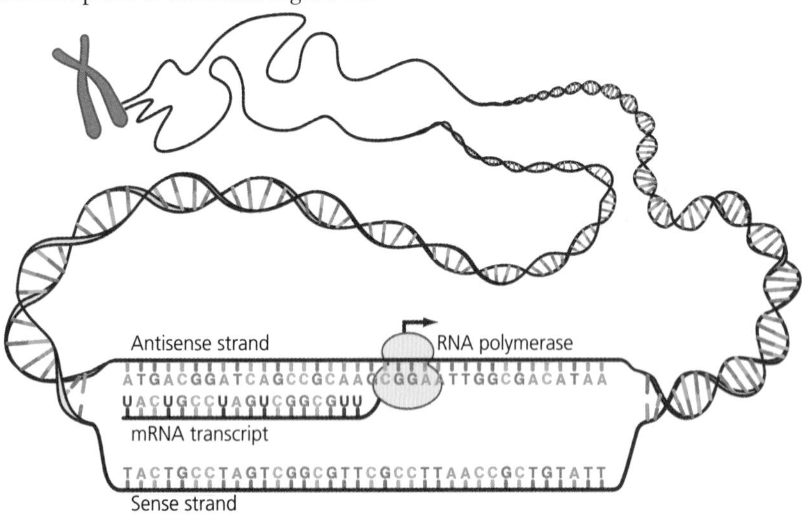

Figure 42 Transcription

At this stage, the mRNA is called pre-mRNA. It contains both introns and exons. The introns are spliced out to produce mature mRNA. This process is shown in Figure 43.

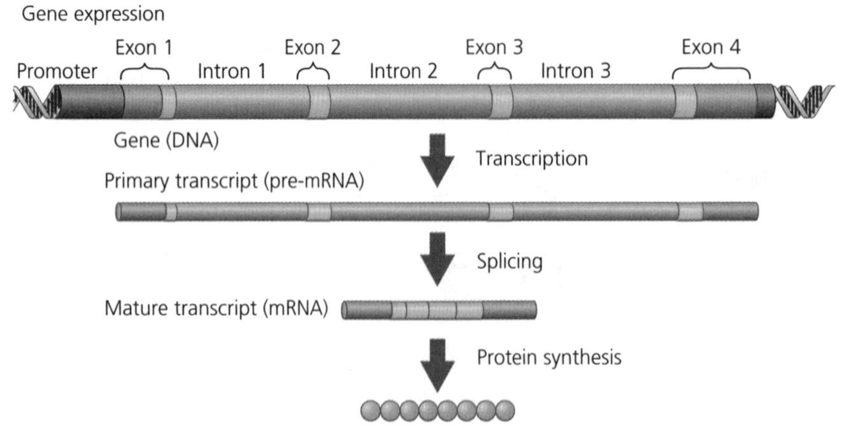

Figure 43 Introns are spliced out to produce mature mRNA

Translation

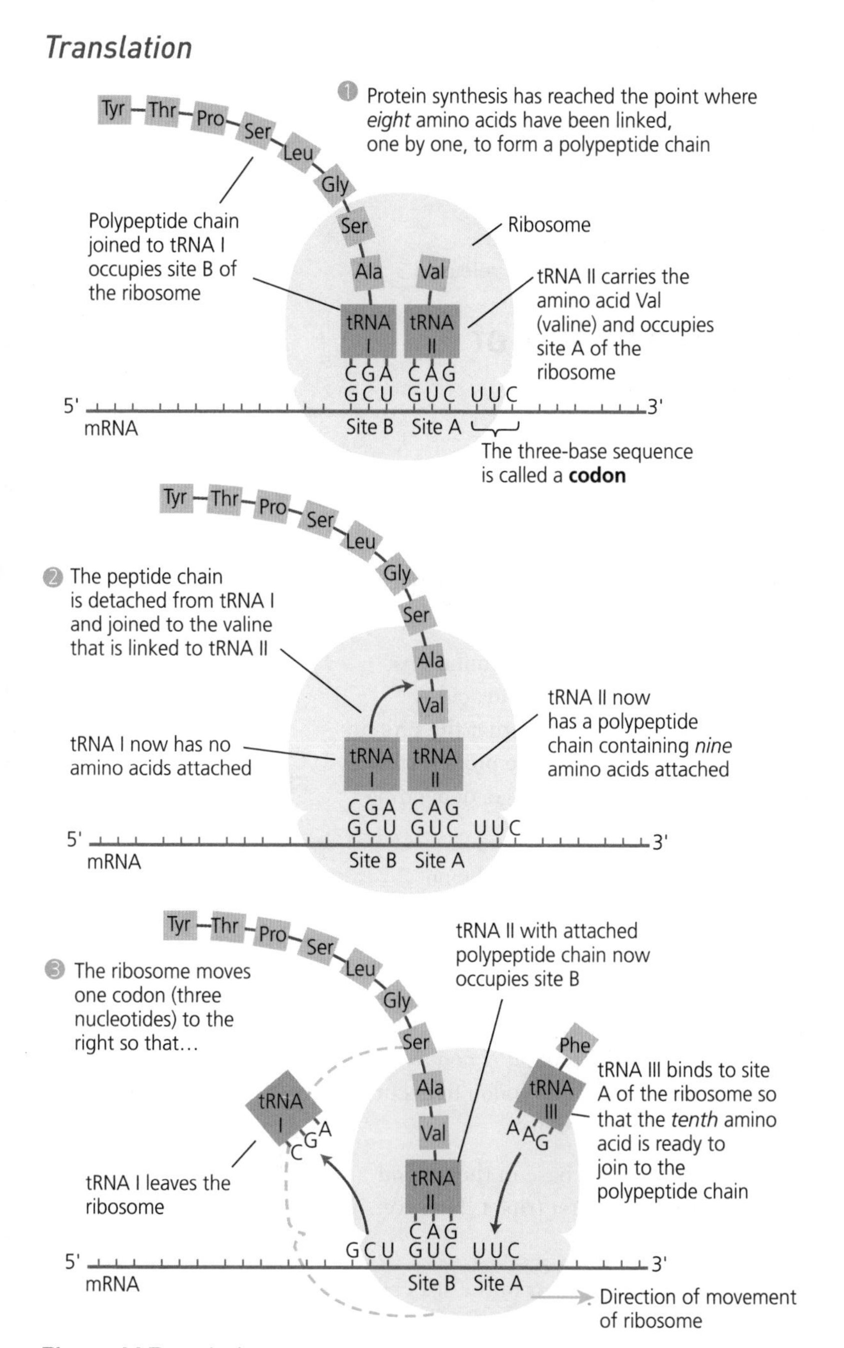

Figure 44 Translation

Translation (Figure 44) is the second stage of protein synthesis:

1. The mature mRNA that has entered the cytoplasm attaches to a ribosome.
2. tRNA molecules with anticodons complementary to the first two codons of the mRNA enter the ribosome, with their amino acids attached.
3. A peptide bond forms between the amino acid molecules. Energy released from ATP is used in this reaction.

Knowledge check 33

If the triplet in the DNA template is ATG, what will the following be?

a the mRNA codon

b the tRNA anticodon

Knowledge check 34

Name the type of reaction that occurs when two amino acids join, forming a peptide bond.

Exam tip

You need to be able to interpret data from experimental work involving the role of nucleic acids, but you do not need to know any specific experiments.

4 The tRNA that bound to the first codon is now free to leave the ribosome.
5 The ribosome moves along the mRNA by one codon and now a new tRNA with an anticodon complementary to the third codon enters the ribosome.
6 A peptide bond forms between the second and third amino acids, again using energy from ATP.
7 The ribosome moves along the mRNA by one codon and the process is repeated until eventually a 'stop' codon is reached and the protein is released.

Genetic diversity can arise as a result of mutation or during meiosis

Mutation

Mutations are changes in the base sequence of chromosomes. They occur most often when DNA is replicating. Two examples are base substitution and base deletion.

In **base substitution**, one base in a DNA triplet is replaced by another, as shown in Figure 45. In this example, thymine is replaced by guanine.

In Figure 45, the triplet ATT has changed to ATG. Like all substitution mutations, this affects only one triplet. The original triplet ATT codes for the amino acid isoleucine, but the new triplet ATG codes for methionine. This means that the protein made by the mutated DNA will have one amino acid different from the protein coded for by the original DNA. This might or might not be a significant change, depending on where in the protein this change has occurred.

A-T-T -T-C-C -G-T-T -A-T-C ...
Original base
A-T-G -T-C-C -G-T-T -A-T-C ...
Substituted base

Figure 45 Base substitution

However, if the original triplet of ATT had mutated to ATC, there would be no change in the protein. This is because DNA is degenerate and both the triplets ATT and ATC code for isoleucine.

In a **base deletion**, a base is missed out. The DNA is still 'read' three bases at a time, so this results in a **frame shift**. Each triplet is affected from the deletion onwards. This results in a very different sequence of proteins, so that the protein is non-functional. Alternatively, the deletion may cause a nonsense or 'stop' codon to occur in the sequence, so no protein is made at all. This is shown in Figure 46.

A-T-T -T-C-C -G-T-T -A-T-C ...
Deletion here
A-T-T -C-C-G -T-T-A -T-C ...
Replaced by first base of next triplet

Figure 46 Base deletion

In Figure 46, the first triplet remains the same because the first base in the second triplet happens to be the same as the deleted third base in the first triplet. However, all the triplets after the deletion are altered.

Mutations are not passed on to the next generation unless they occur in the sex cells or in the cells that produce the sex cells.

The rate at which mutations occur can be increased by mutagenic agents such as:

- toxic chemicals, e.g. mustard gas, bromine compounds or nitrosamines found in tobacco smoke
- ionising radiation, e.g. X-rays or gamma rays
- high energy radiation, e.g. ultraviolet light.

Exam tip

If you are asked to name a mutagenic agent, give a specific example. Don't just say 'chemicals' or 'radiation' as this implies that all kinds of chemicals or radiation can cause mutations, which is not true.

Meiosis

Meiosis is a specialised kind of cell division that produces daughter cells with:

- half the number of chromosomes as the parent cell; in animals and flowering plants, meiosis produces gametes
- different combinations of alleles from each other and from the parent cell

In animals such as humans, body cells are diploid which means that they contain two sets of chromosomes. Meiosis is a two-stage division that produces four daughter cells which are all haploid, containing one set of chromosomes.

Before meiosis (just as in mitosis), DNA replication occurs. However, unlike mitosis, in the early stages of meiosis the homologous chromosomes pair up. The chromosomes line up side by side, and their chromatids cross over at points called **chiasmata** (*singular*: chiasma). You can see this in Figure 47.

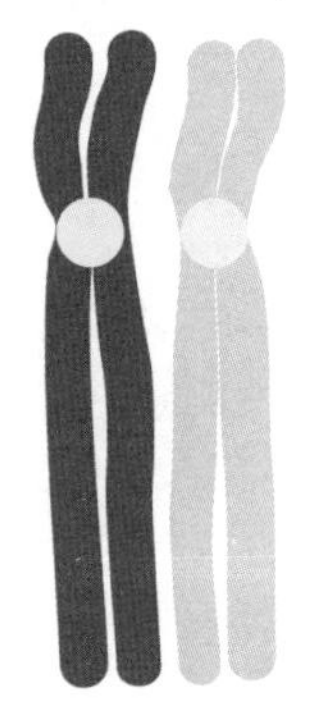

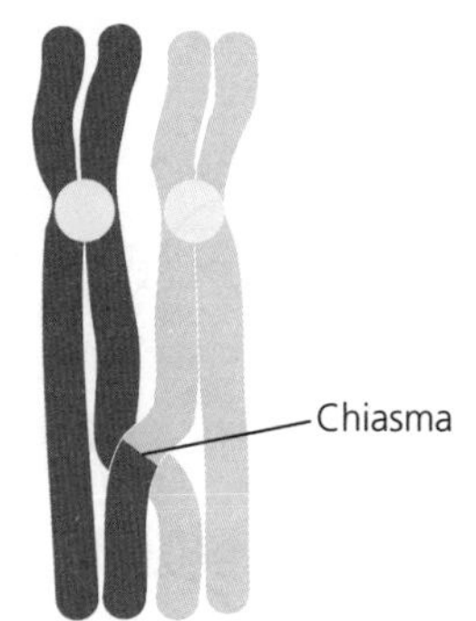

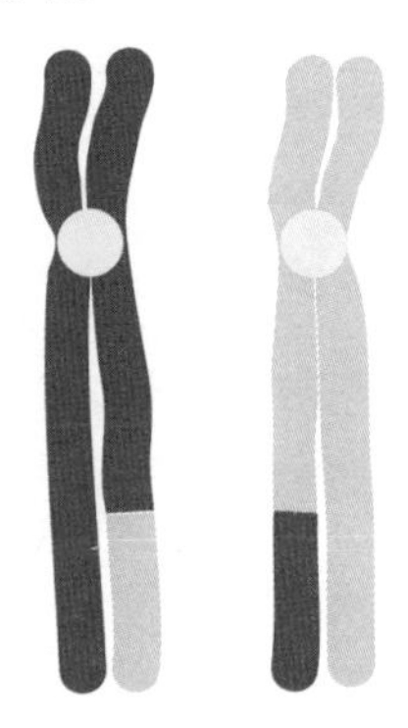

Figure 47 Crossing over of chromatids of homologous chromosomes

Another feature of meiosis that results in genetic variation is **independent assortment**. During the first division of meiosis, the pairs of homologous chromosomes line up on the equator before the spindle fibres pull them to opposite poles of the cell. Each pair behaves independently of every other pair, so many possible different combinations can result. Again in the second division, the chromatids move to opposite poles of the cell independently of every other pair. This also results in the shuffling of alleles. You can see this in Figure 48.

There are 2^n possible combinations that can result from independent assortment alone, where n is the number of chromosomes. In a human, there are 2^{23} combinations, or 8 338 608. However, when crossing-over is taken into account, a vast number of different genetic combinations is possible in the gametes from just one individual.

Mutations in the number of chromosomes can occur during meiosis. This is called **non-disjunction**. This can result in one daughter cell having two copies of a chromosome and the other daughter cell having a missing chromosome. When the gamete is fertilised, three copies of a chromosome may be present in the zygote or the zygote may have only one copy of a chromosome instead of two.

Knowledge check 35

Explain the difference between a homologous pair of chromosomes and a chromosome made up of chromatids.

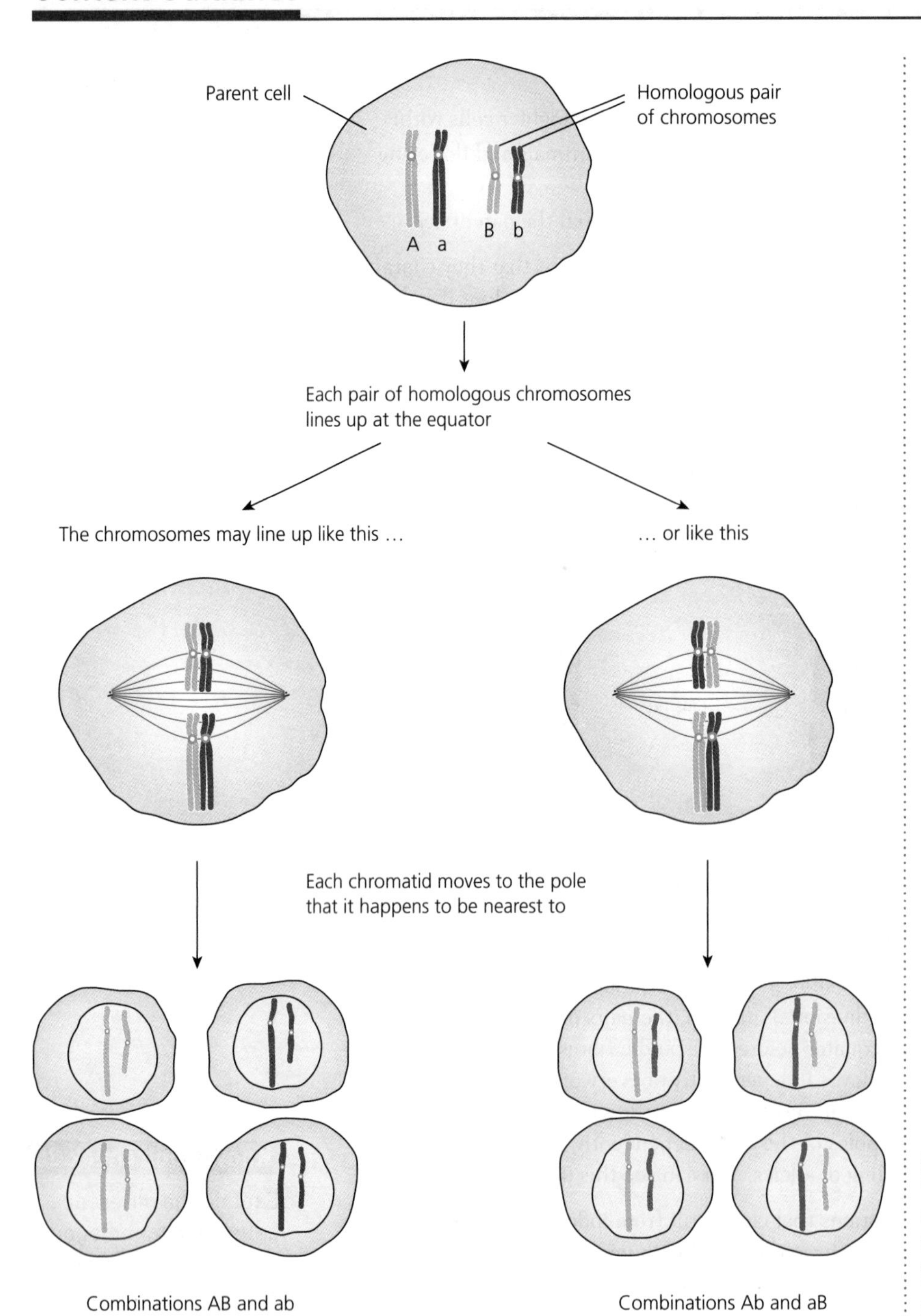

Figure 48 Independent assortment

Another source of variation is **random fertilisation**. Usually, there are many male gametes and there may be several female gametes. It is random which male gamete fertilises which female gamete.

Exam tip

Although random fertilisation is an important source of variation, it does not happen during meiosis so it is not a way in which meiosis contributes to variation.

Knowledge check 36

The diagram shows the life cycle of a moss, which has both haploid and diploid stages in its life-cycle.

Annotate the diagram with:

- F at a place where fertilisation takes place
- X at a place where mitosis takes place
- M at a place where meiosis takes place

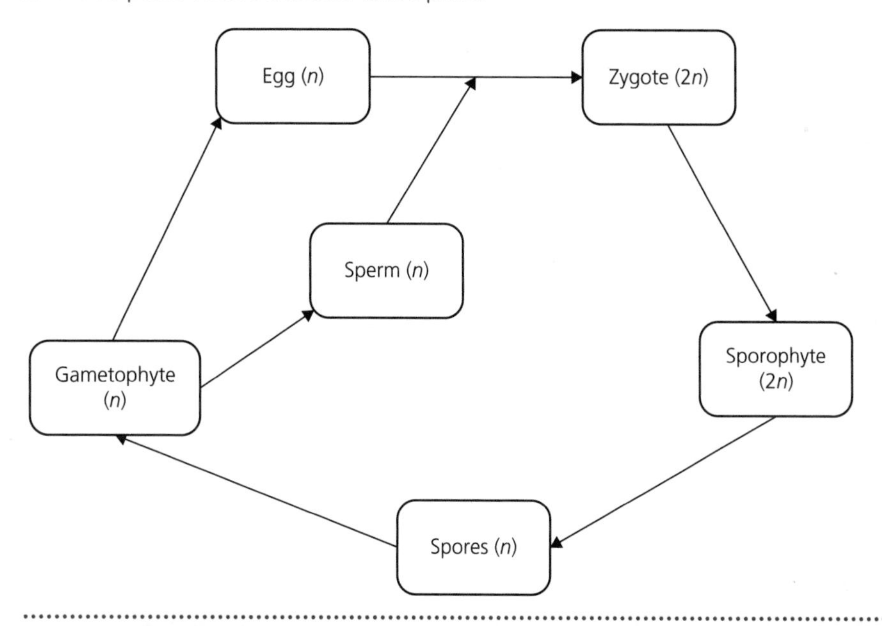

Summary

- DNA in prokaryotes is circular and not complexed with proteins. DNA in eukarotes is longer, linear and wound around proteins to form chromosomes. Eukaryotic DNA contains non-coding DNA including short repeated sequences and introns.
- Protein synthesis involves transcription. This occurs in the nucleus. The base sequence of the DNA in a gene is copied into a molecule of mRNA.
- In translation, tRNA molecules bring specific amino acids to the ribosome. tRNA binds to the codon in mRNA by complementary base pairing. The amino acids join by a peptide bond.
- Mutations may occur, especially during DNA replication. Two examples are substitution and deletion.
- Meiosis is a specialised form of cell division that results in the formation of four haploid daughter cells that are all genetically different.
- Genetic variation in the daughter cells results from crossing-over of chromatids and independent assortment.

Genetic diversity and adaptation

Genetic diversity is the number of different alleles in a population. Mutation occurs at random and, although most mutations result in non-functional proteins, some result in new alleles that may be advantageous, or at least not disadvantageous. It is an important factor in natural selection. Natural selection leads to the evolution of populations:

- Random mutation results in a new allele, which is advantageous to the organism that possesses it.
- This individual survives longer and therefore produces more offspring than many other members of the population.
- These offspring inherit the advantageous allele and also survive longer. They produce more offspring, which also inherit the advantageous allele.
- Over many generations, this allele increases in frequency within the population.

Directional selection is when an advantageous allele becomes more frequent in a population and **stabilising selection** is when an unfavourable allele becomes less frequent. An example of directional selection is the development of antibiotic resistance in bacteria, as shown in Figure 49.

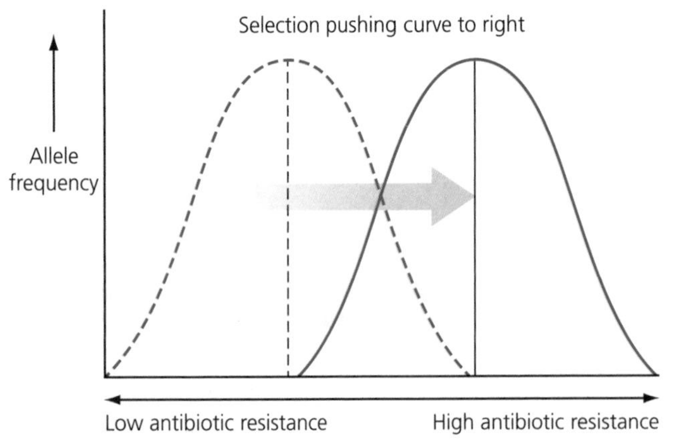

Figure 49 Directional selection increases antibiotic resistance

Figure 50 explains how this occurs. This example is about a population of the bacteria that cause TB in the body of a TB patient. At the start, none of the bacteria are resistant to the antibiotic isoniazid. By chance, one of the bacteria undergoes a mutation, resulting in resistance to isoniazid. All the other bacteria are killed by the antibiotic, but the resistant bacterium survives and passes on the allele for resistance to its offspring. Now all the bacteria in the TB patient are resistant to isoniazid.

Exam tip

Remember that bacteria become resistant to antibiotics — they do not become immune. Immunity involves the production of antibodies and memory cells. These bacteria are simply resisting the antibiotic.

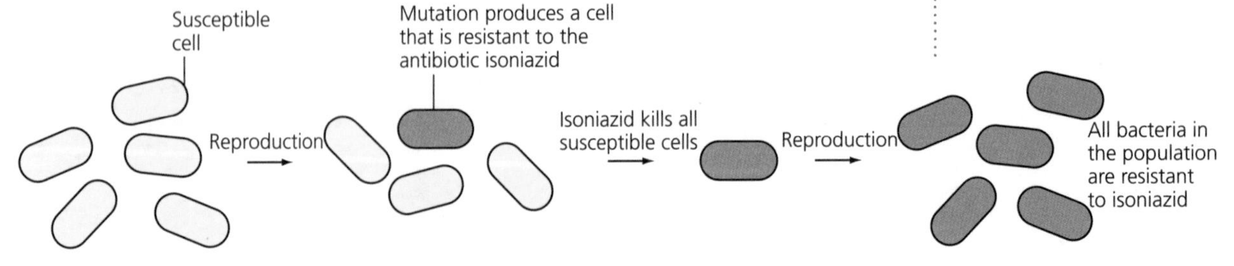

Figure 50 The evolution of antibiotic-resistant populations of bacteria

Figure 51 shows stabilising selection. An example of this is birth mass in humans. Infants born with very low or very high birth masses are less likely to survive, so the alleles that determine an intermediate or optimal birth mass are selected for.

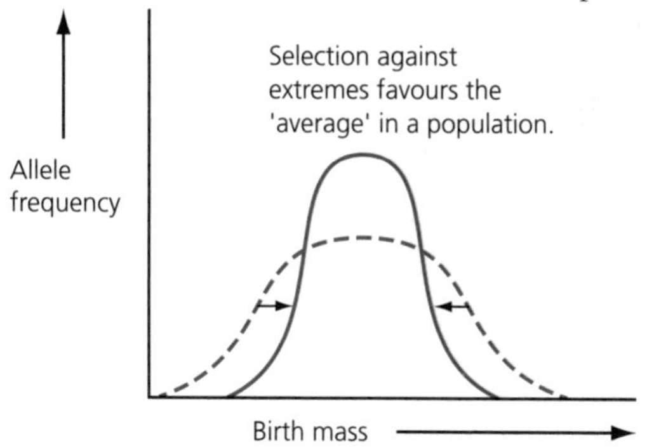

Figure 51 Stabilising selection acts to maintain optimal human birth mass

Knowledge check 38

The number of eggs laid at one time by female birds (clutch size) is genetically determined. Scientists studying a particular species of bird found that most females of that species laid four or five eggs at a time. Few birds produced larger or smaller clutch sizes than this.

a What type of selection is occurring?

b Explain the advantage of this to this species of bird.

Knowledge check 37

Scientists have studied the DNA of ancient bacteria, thousands of years old, found in permafrost sediments. They have found genes that are resistant to antibiotics in this DNA. Suggest how this has occurred.

Required practical 6

Use of aseptic techniques to investigate the effect of antimicrobial substances on microbial growth

In hospital laboratories, tests may be carried out to find the most effective antibiotic to prescribe for a patient with a bacterial infection. In this case, bacteria taken from a swab from the patient are spread on a sterile Petri dish containing an agar medium on which the bacteria will grow. Paper discs impregnated with antibiotics are placed on the agar surface and the dish is incubated at 37°C for 24 hours. After this, clear zones appear around some of the paper discs where the bacteria have not been able to grow. These are called **inhibition zones**. The larger the inhibition zone, the more effective that antibiotic is at killing the bacteria.

In schools and colleges, it is unsafe to use bacteria from a swab, even if the person is healthy. A student simulated the laboratory test described above by using a broth culture of a non-pathogenic bacterium obtained from a reputable source. She used a sterile pipette to transfer two drops of the broth culture to the surface of some sterile nutrient agar in a Petri dish. Next, she used a sterile spreader to spread the bacteria over the whole of the surface of the agar. After this, she used sterile forceps to place a mast ring (a paper disc impregnated with known concentrations of several different bacteria) on the surface of the agar. She sealed the lid on the dish with two 'hinges' of Sellotape and incubated the dish for 24 hours. Figure 52 shows the dish as it appeared after incubation.

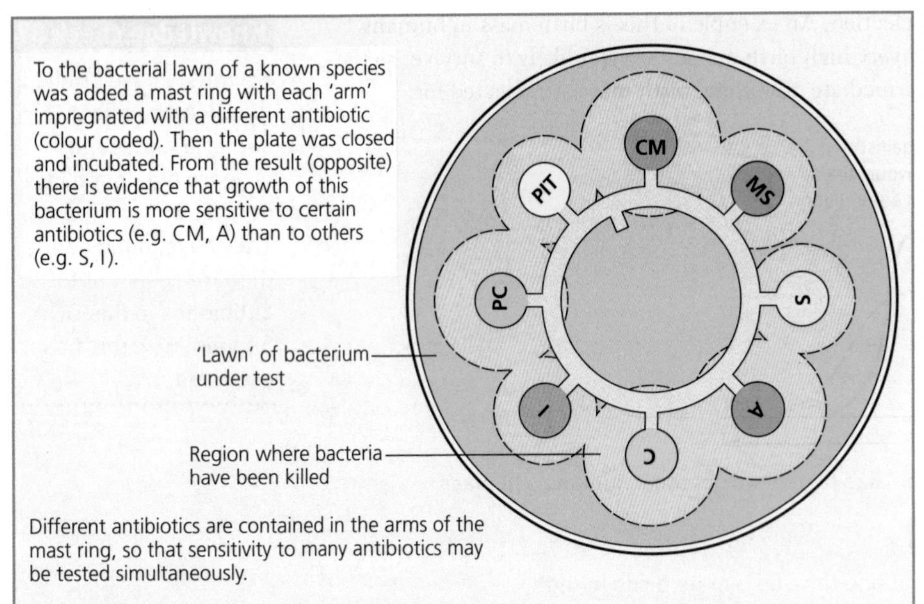

Figure 52 Investigating sensitivity to antibiotics

1. The student used aseptic technique throughout. Give two reasons why this is necessary.
2. Why is it considered unsafe for school or college students to use bacteria from a swab in this investigation, even if the swab is taken from a healthy person?
3. What measurements could the student make to obtain quantitative results?
4. If these were bacteria from a patient with a bacterial infection, use the student's results to suggest the antibiotic a doctor should prescribe. Explain the reason for your answer.
5. Suggest two reasons why a doctor might prescribe a different antibiotic from the one(s) you suggested in question 4.

Turn to page 91 for the answers.

Species and taxonomy

Two organisms belong to the same species if they are able to interbreed and produce fertile offspring. In many species such as birds, males and females can look very different. Males may have brightly coloured plumage, whereas the females of many species are a dull, brown colour. One way in which a bird can recognise a potential mate of the correct species is by **courtship behaviour**. This is an innate behaviour that is specific to one species. An example for a fish, the three-spined stickleback, is shown in Figure 53.

Every species is given a **binomial name**, consisting of its genus and species name. For example, humans are *Homo sapiens*. Note that the genus starts with a capital letter and the species starts with a lower case letter. The whole name is written in italics or underlined. Humans are the only living members of the genus *Homo*, but

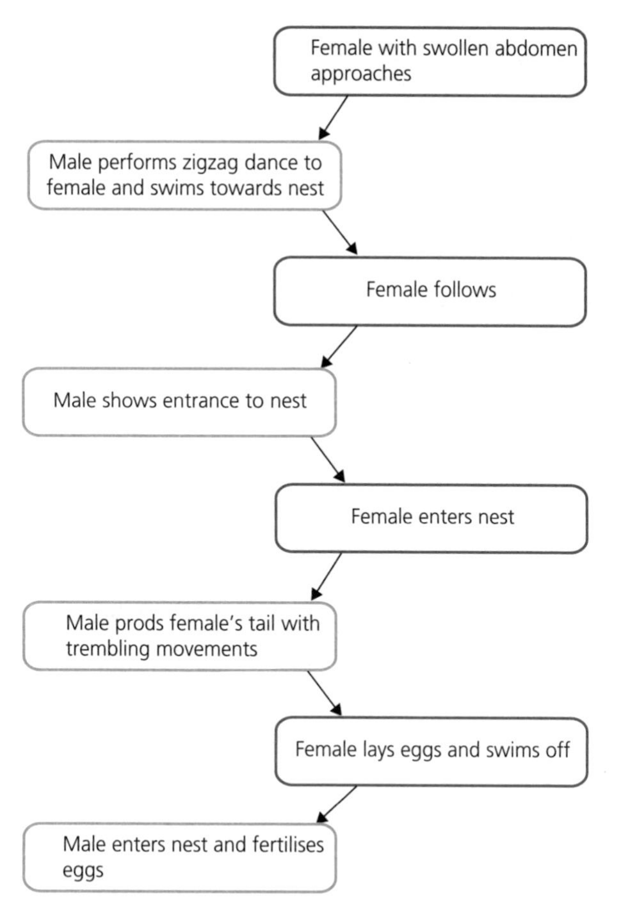

Figure 53 Courtship behaviour of the three-spined stickleback

in the past there were other species of humans including *Homo habilis* and *Homo erectus*. The fact that these species share the same genus but not the same species name tells us that, although they are different species, they are closely related to modern humans.

Living organisms are classified into groups based on their evolutionary origins and relationships. This is called a **phylogenetic system**. Each group is called a **taxon** (*plural:* taxa). These are arranged in a **hierarchy**, which means that large groups are subdivided into smaller groups, with no overlap. Table 1 shows how modern humans are classified. As you go down the hierarchy, there are fewer organisms in each taxon.

Scientists can use different kinds of evidence to work out the evolutionary relationships between different organisms. One way is to compare their **genome sequences**. This means that we can find the base sequence of the DNA of the two organisms and find out how many differences there are. The fewer the number of differences, the more closely related the organisms are. There are other, more indirect, ways of comparing the genome sequence. For example, the mRNA of two organisms can be compared. As DNA codes for proteins, counting the number of differences in the amino acid sequence of important proteins, such as haemoglobin, can also be used.

Another source of evidence is **immunology**. One key protein found in many organisms is the blood protein, albumin. Figure 54 shows how immunology can be used to investigate how closely related are humans, spider monkeys and pigs.

Knowledge check 39

Apart from species recognition, suggest two other functions of courtship behaviour.

Exam tip

You don't need to know any specific examples of courtship behaviour, but you may be given an example and be asked questions about it.

Knowledge check 40

Three ancestors of modern humans are *Homo erectus*, *Homo habilis* and *Australopithecus africanus*. What do these names tell you about their evolutionary relationship?

Taxon	Humans as an example	Explanation
Domain	Eukarya	This group includes all organisms with eukaryotic cells
Kingdom	Animalia	Humans are multicellular organisms. Their cells have no cell walls
Phylum	Chordata	Humans are animals with backbones
Class	Mammalia	Humans have hairy skin and produce milk to feed their young
Order	Primates	Primates have flat faces, forward-facing eyes and opposable digits
Family	Hominidae	These are human-like creatures, including humans and great apes such as chimpanzees and gorillas
Genus	*Homo*	Modern humans are the only living members of this genus
Species	*sapiens*	All modern humans belong to this species

Table 1 The classification of modern humans

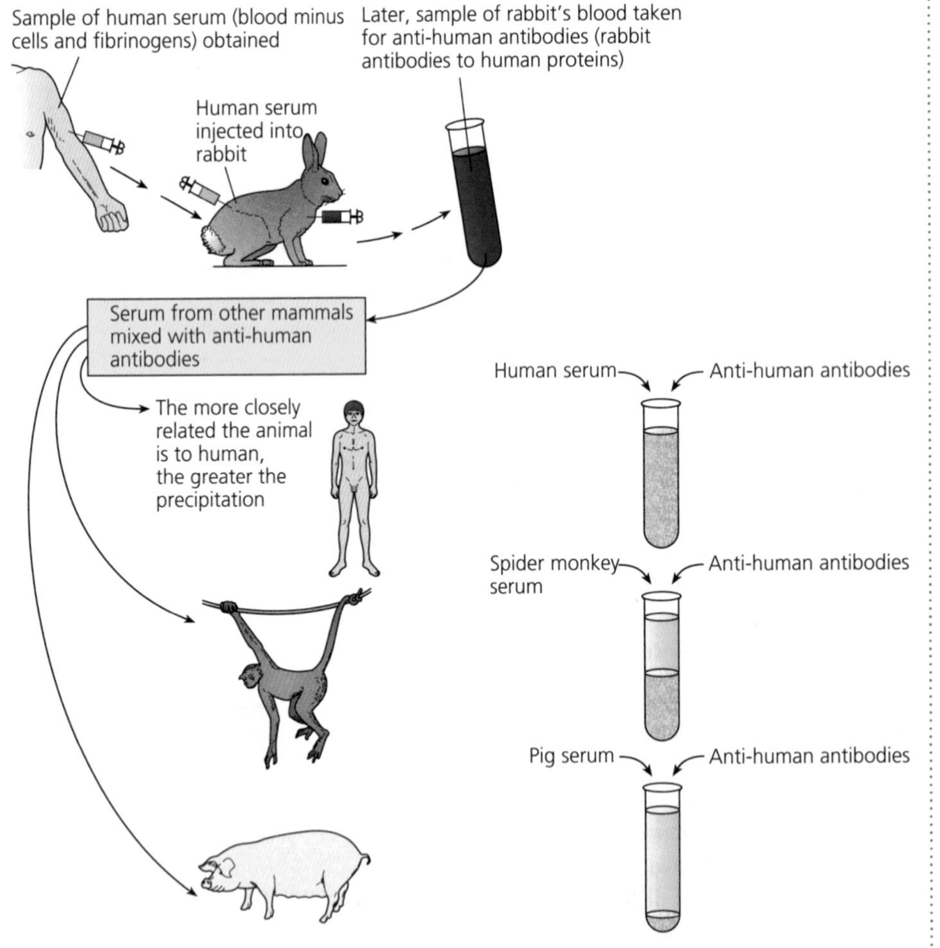

Figure 54 The immune reaction and evolutionary relationships

Exam tip

You do not need to know the classification of any organism, just the names of the taxa shown.

Exam tip

You can use a mnemonic to remember the hierarchy in Table 1. For example, Daring King Philip Came Over For Green Spinach will give you the initial letter of these taxa in the right order.

Exam tip

Immunological studies are a means of detecting differences in specific proteins of species, and therefore (indirectly) their **relatedness**.

Knowledge check 41

Scientists injected human blood serum into rabbits, which the rabbits formed antibodies against. Rabbit antibodies were mixed with human blood and a precipitate was formed. The scientists then mixed the rabbit antibodies against human blood serum with samples of blood serum from other animals. They compared the volume of the precipitate formed with the volume of precipitate formed when the rabbit antibodies were mixed with human blood serum. The results were as follows.

Source of blood serum	Volume of precipitate/%
Human	100
Chimpanzee	97
Baboon	50
Dog	0

What do these results tell you about the evolutionary relationship between these four organisms?

Biodiversity within a community

Biodiversity is a measurement of the variety of living organisms present in different ecosystems. This includes genetic variation and the number of species within an area. Biodiversity can relate to a range of habitats ranging from small local habitats to the Earth.

Species richness is a measure of the number of different species in a community. An **index of diversity** measures not only the number of species in a community, but also the number of individuals in each species. One formula for the index of diversity (d) can be calculated using the formula:

$$d = \frac{N(N-1)}{\Sigma n(n-1)}$$

where N is the total number of organisms of all species and n is the total number of organisms of each species.

Σ = the sum of

Knowledge check 42

A student counted the number of birds of different species that visited his garden in a half-hour period. The results are shown in the table.

Species	Number
House sparrow	3
Thrush	4
Blue tit	7
Pigeon	2
Magpie	4
Blackbird	5

a Calculate the species richness and the species diversity index for birds in this garden.

b Evaluate the use of species richness and a species diversity index in assessing biodiversity in an area.

Farming techniques reduce biodiversity. Some of the reasons for this include the fact that farmers:

- use pesticides that kill species other than the crop they are growing
- plough fields and plant large areas with plants of the same species
- remove hedges to produce larger fields where large machines can be used more easily
- drain fields and add fertiliser to maximise crop production on the land

All these activities increase food production — it is important to provide a large population with enough food at an affordable price. However, there are conservation measures that farmers can take to improve biodiversity. In return for carrying out these measures, farmers can attract a higher price for their crops. Examples of conservation measures include:

- growing wildflowers and clover in field margins to provide nectar for insect food and habitats for insects
- growing a proportion of plants that provide seeds for birds in winter and early spring
- growing tussocky and fine grasses that provide a habitat for spiders, beetles and small mammals, which in turn provide food for predators such as barn owls
- developing features such as woodland, old barns, ponds, hedges and ditches as wildlife areas

Exam tip

You do not need to know any specific conservation schemes, but you will be expected to evaluate information you are given about conservation measures, including data.

Investigating diversity

The genetic diversity within, or between, species can be made by comparing:

- the frequency of measurable or observable characteristics
- the base sequence of DNA
- the base sequence of mRNA
- the amino acid sequence of the proteins that DNA and mRNA code for

However, now that sequencing DNA is quicker and more reliable, measuring the frequency of observable characteristics is increasingly being replaced by comparison of DNA sequences. When comparing mRNA or DNA sequences, the more differences that are present, the greater the genetic diversity.

Exam tip

This part of the specification is likely to be tested by giving you data to interpret. Therefore, it is important that you understand the underlying principles. Remember that if there is low genetic diversity within a species, it means that the species is less likely to survive if the environment changes. You need to link this in with what you know about natural selection.

Knowledge check 43

The table shows the number of amino acid differences between one haemoglobin polypeptide in a human and some other animals. What do these data tell you about the genetic variation between these different species?

Animal	Number of amino acid differences
Human	0
Bird	45
Dog	32
Frog	67
Lamprey	125
Macaque	8

Investigating variation within a species

When investigating variation within a population, it is important that we obtain a sample that is large enough to be representative of the whole population. It is also important that we sample at **random**. In a random sample, every individual in the population has an equal chance of being included in the sample.

If you wanted to take a random sample of humans, for example to measure their height, you would find a large population and then pull names out of a hat to select a smaller random sample of, say, 50. Another way to do this would be to give everyone a number and then to use a random number generator to select a random sample.

To sample plants or animals that do not move (e.g. limpets on rocks), you should place several **quadrat frames** at random in the area. Most quadrat frames are a metal square, often 0.5 m by 0.5 m. To place the quadrat at random, you need to divide the area to be sampled into a grid. Then you use a random number generator on a computer or a calculator to give you a pair of random numbers. These are used as x and y coordinates. You place the quadrat with its bottom left-hand corner at this point and then measure all of the organisms of the species you are studying within this quadrat. This is repeated several times (at least ten). You can see this in Figure 55. In this figure, the x coordinate was 5 and the y coordinate was 4.

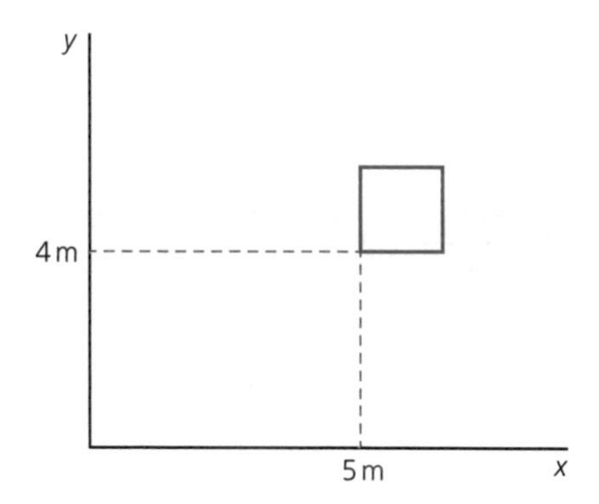

Figure 55 Placing a quadrat frame at random

Mobile organisms such as birds, small mammals or insects can be sampled using small mammal traps, sweep nets etc.

When you have sampled the population, you normally calculate a mean value. This means that you add up all of the measurements you took (e.g. height in metres) and then divide this total by the number of individuals in the sample.

However, calculating a mean does not tell you anything about the variability of the data. To do this, standard deviation is used. This gives a spread within which 68% of values fall (34% above the mean and 34% below the mean). Put simply, a high figure for standard deviation means that the measurements in the sample are varied, and a low figure for standard deviation means that the measurements in the sample are not very varied and most of them are close to the mean value. Standard deviation values are useful when comparing the data from two samples.

Exam tip

You will not be asked to calculate standard deviation in written papers, but you do need to understand standard deviation values when interpreting data.

In Figure 56 you can see the data collected for shell length in two different species of snail. The bar chart shows the mean length of shell for each species, but there are also lines included above and below the means. These lines show the standard deviation for the two samples. A careful look at these bars will show you that the standard deviation bars do not overlap between the two populations, although they are close. This means that we can be fairly certain that there is a statistically significant difference between the shell lengths of these two populations. On the other hand, if these bars did overlap, we could not be sure that there is a statistically significant difference between the two populations.

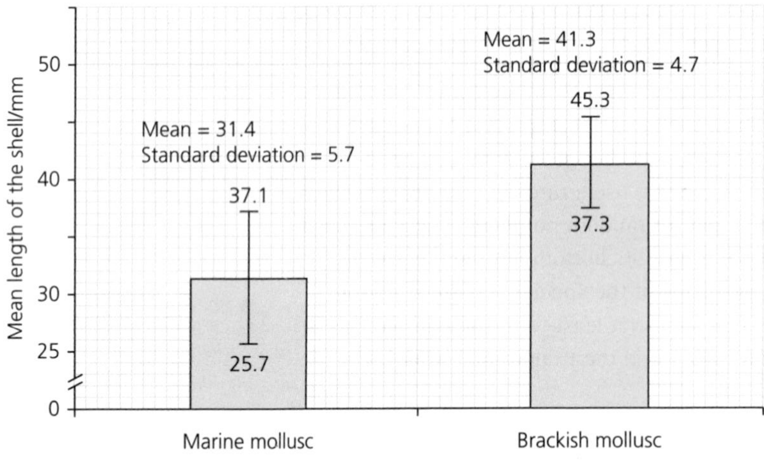

Figure 56 Shell lengths in molluscs

Summary

- Two kinds of natural selection are directional selection, as shown by antibiotic resistance in bacteria, and stabilising selection, as shown by human birth mass.
- A species is a group of organisms that can interbreed to produce fertile offspring.
- Each species has a binomial name. Organisms can be classified in a phylogenetic hierarchy that reflects their evolutionary relationships.
- It is possible to investigate how closely related organisms are by comparing their DNA base sequences or the structure of their proteins, using immunology.
- Biodiversity can be measured by calculating species richness or using an index of diversity.
- Many farming practices reduce biodiversity, although there are conservation measures that can be carried out by farmers.
- When measuring variation in a population, different sampling methods are used depending on the type of organism being studied.

Questions & Answers

Exam format

If you are taking AS biology, your exams will be structured as follows:

Paper 1	Paper 2
Any content from topics 1–4, including relevant practical skills	Any content from topics 1–4, including relevant practical skills
Written exam, 1 hour 30 minutes 75 marks, worth 50% of AS	Written exam, 1 hour 30 minutes 75 marks, worth 50% of AS
65 marks: short-answer questions 10 marks: comprehension question	65 marks: short-answer questions 10 marks: extended response question

If you are taking A-level biology, your exams will be structured as follows:

Paper 1	Paper 2	Paper 3
Any content from topics 1–4, including relevant practical skills	Any content from topics 5–8, including relevant practical skills	Any content from topics 1–8, including relevant practical skills
Written exam, 2 hours 91 marks, worth 35% of A-level	Written exam, 2 hours 91 marks, worth 35% of A-level	Written exam, 2 hours 78 marks, worth 30% of A-level
76 marks: mixture of long- and short-answer questions 15 marks: extended response	76 marks: mixture of long- and short-answer questions 15 marks: comprehension	38 marks: structured questions, including practical techniques 15 marks: critical analysis of experimental data 25 marks: essay from a choice of two titles

Tips for answering questions

Use the mark allocation. Generally, one mark is allocated for one fact, concept or item in an explanation. Make sure your answer reflects the number of marks available.

Respond appropriately to the command words in each question, i.e. the verb the examiner uses. The terms most commonly used are explained below.

- **Describe** — this means 'tell me about…' or, sometimes, 'turn the pattern shown in the diagram/graph/table into words'; you should not give an explanation.
- **Explain** — give biological reasons for *why* or *how* something is happening.
- **Calculate** — add, subtract, multiply, divide (do some kind of sum!) and show how you got your answer — *always* show your working!
- **Compare** — give similarities *and* differences between…

- **Complete** — add to a diagram, graph, flowchart or table.
- **Name** — give the name of a structure/molecule/organism etc.
- **Suggest** — give a plausible biological explanation for something; this term is often used when testing understanding of concepts in an unfamiliar context.
- **Use** — you must find and include in your answer relevant information from the passage/diagram/graph/table or other form of data.

About this section

This section contains questions similar in style to those you can expect to see in your exam. The limited number of questions in this guide means that it is impossible to cover all the topics and all the question styles, but they should give you a flavour of what to expect. The responses that are shown are students' answers to the questions.

The papers have the same number of marks as AS paper 1 (Test paper 1) and A-level paper 1 (Test paper 2). Test paper 1 is similar in style to AS paper 1 and Test paper 2 is similar in style to A-level paper 1. There is overlap in the content covered in Test paper 2 and the AS paper, so Test paper 2 can be used if you are studying AS for extra practice. No questions typical of paper 3 have been included here because they are synoptic, and you haven't covered enough topics at this stage.

There are several ways of using this section. You could:

- hide the answers to each question and try the question yourself. It needn't be a memory test — use your notes to see if you can actually make all the points you ought to make
- check your answers against the students' responses and make an estimate of the likely standard of your response to each question
- check your answers against the examiner's comments to see where you might have failed to gain marks
- check your answers against the terms used in the question — for example, did you *explain* when you were asked to, or did you merely *describe*?

Examiner's comments

Each question is followed by a brief analysis of what to watch out for when answering the question (icon ⓔ). Student responses are then followed by detailed comments. These are preceded by the icon ⓔ and indicate where credit is due. In the weaker answers, they also point out areas for improvement, specific problems and common errors, such as lack of clarity, weak or non-existent development, irrelevance, misinterpretation of the question and mistaken meanings of terms.

Test paper 1

This paper has 75 marks available and should take 1 hour 30 minutes to complete. It is similar in style to AS paper 1.

Question 1

(a) Give two ways in which meiosis contributes to variation. (2 marks)

Figure 1 shows a cell at the start of meiosis.

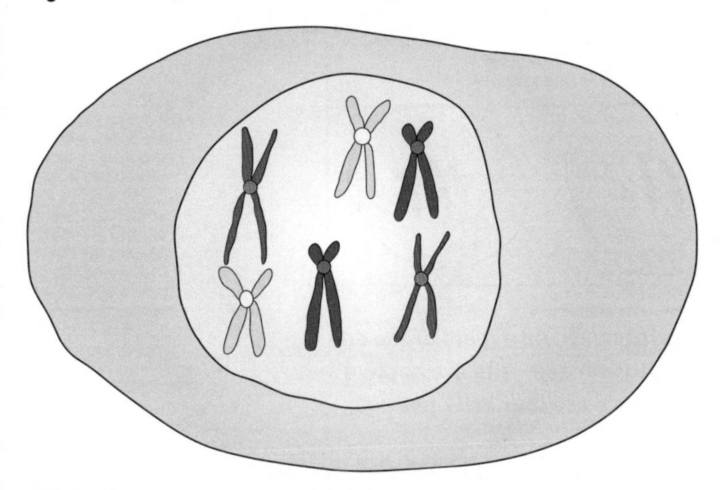

Figure 1

(b) In the space below, sketch the appearance of the daughter cells produced at the end of the first division of meiosis. (3 marks)

e This question starts with an easy recall section. Part (b) is straightforward too, as long as you remember it is asking about the end of the first meiotic division.

Student A

(a) Independent segregation ✓ and crossing-over ✓.

e **2/2 marks awarded** These are both right, so 2 marks are gained.

(b)

e **3/3 marks awarded** Student A gets all 3 marks: the cells are haploid and consist of chromosomes made of two chromatids, and there are two daughter cells.

Student B

(a) One way in which meiosis contributes to variation is crossing-over ✓ and another is random fertilisation.

e 1/2 marks awarded Student B gains 1 mark for 'crossing-over', but not for 'random fertilisation'. Although random fertilisation is a source of variation, it doesn't happen during meiosis. Note that the student could have just written 'crossing-over' and saved time by not writing the full sentence starting 'One way in which meiosis contributes to variation is...'

(b)

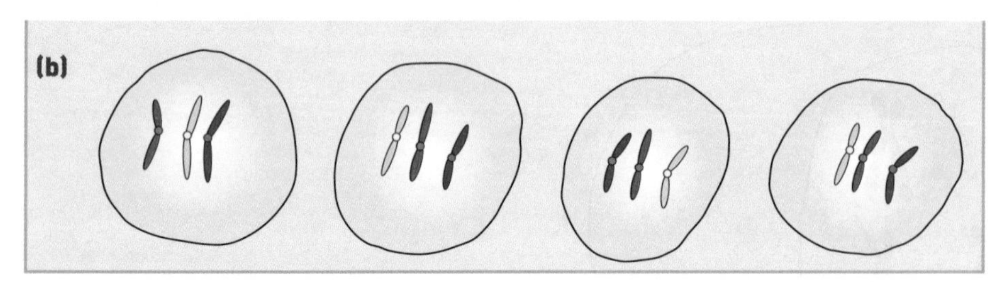

e 1/3 marks awarded Student B has drawn the cells you would expect at the end of the second division of meiosis. 1 mark is given because these cells are haploid and there is clearly one of each pair in each cell, but there are four cells not two, and the chromosomes do not consist of pairs of chromatids.

Question 2

The apparatus in Figure 2 was used in an investigation into gas exchange in a grasshopper (an insect). The scientists showed that the grasshopper took in air through the first four pairs of spiracles and expelled air through the last six pairs of spiracles. The rubber diaphragm prevents air flow except through the grasshopper.

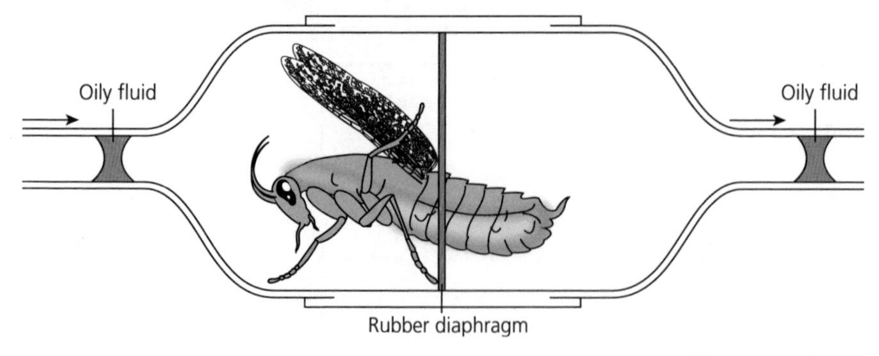

Figure 2

(a) (i) **Explain how the scientists could use the apparatus to show that there is one-way flow of air through the tracheal system of the grasshopper.** (2 marks)

(ii) **Explain the advantage to the grasshopper of one-way flow of air in the tracheal system.** (1 mark)

(b) What measurements would you need to take to find the volume of air taken in by the grasshopper per gram of body mass per minute? (2 marks)

Figure 3 shows a mosquito larva. This is an insect that lives under the surface of freshwater ponds.

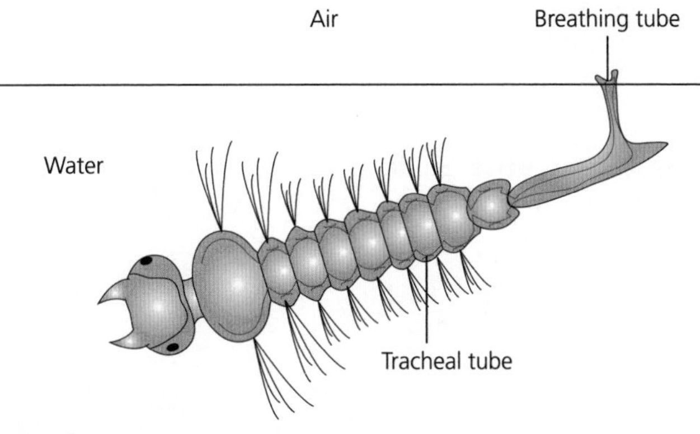

Figure 3

(c) Some aquatic insects have gills that take in oxygen dissolved in the water. Suggest an advantage to the mosquito larva of obtaining oxygen from the air. (1 mark)

e This question asks you to apply your knowledge to a new situation, although you are not expected to be familiar with this investigation. You should be able to answer these questions if you understand what you have learned about gas exchange in insects.

Student A

(a) (i) The bubble of oily fluid at the front of the tube will move towards the grasshopper as air is taken in, and the bubble of oily fluid at the back of the tube will move away from the grasshopper ✓ as air is released through the rear spiracles ✓.

e 2/2 marks awarded This gains both marks as we are told which way the oily fluid will move and there is an explanation for the direction of movement.

(ii) The air that is high in oxygen doesn't mix with stale air (low in oxygen) coming out, so this maintains a high concentration gradient ✓.

e 1/1 mark awarded This gets the mark.

(b) The distance moved by the bubble at the front of the tube in a measured time and the radius of the capillary tube so you can find the volume of air taken up ✓. Then the grasshopper must be weighed in grams ✓.

e 2/2 marks awarded This gains both marks as this will give the volume of air taken up in a specific time per gram of grasshopper.

(c) There is a higher % of oxygen in air than in water, so it is a more efficient ✓ way for the mosquito larva to obtain all the oxygen it needs.

ⓔ **1/1 mark awarded** This is a clear advantage and gains the mark.

Student B

(a) (i) Watch the bubbles and they will both move.

ⓔ **0/2 marks awarded** This doesn't gain any marks as it is imprecise. Student B doesn't say which way the bubbles will move and 'watch' doesn't imply any measurement.

(ii) This means fresh air, high in oxygen, keeps entering the system and it does not mix with stale air ✓.

ⓔ **1/1 mark awarded** This is worth the mark, but is not such a good answer as student A gave.

(b) The distance moved by the left-hand bubble in a minute and the mass of the grasshopper ✓.

ⓔ **1/2 marks awarded** Student B gets 1 mark for the mass of the grasshopper and the distance moved by the bubble in a specific time, but the distance moved does not represent a volume and so the second mark is not awarded.

(c) This means that it doesn't need gills.

ⓔ **0/1 mark awarded** This doesn't gain the mark as it is not clear why breathing air is an advantage.

Question 3

(a) What is digestion? (2 marks)

(b) (i) Describe how lipids are digested and absorbed in the gut. (4 marks)

(ii) People who have had their gall bladder removed are advised to take a bile salt supplement before every meal that contains fats. Explain why. (2 marks)

ⓔ Part (a) asks for a simple definition that you should know and (b) (i) is also straightforward recall of what you should have learned. Part (b) (ii) tests your understanding of the topic by giving you a situation you are unlikely to be familiar with.

Student A

(a) The hydrolysis of large molecules into smaller soluble molecules ✓ that can be absorbed ✓.

e 2/2 marks awarded Both marks awarded for a fully correct and well-expressed answer.

(b) (i) Bile from the liver emulsifies fats, increasing the surface area for lipase to act on ✓. Lipase enzymes from the pancreas hydrolyse lipids into fatty acids, glycerol and monoglycerides ✓. Micelles form, made from long chain fatty acids, phospholipids and glycerol ✓. They break down to release monoglycerides and fatty acids which diffuse into the epithelial cells ✓. Inside the epithelial cells, triglycerides are reformed and packed into chylomicrons, which enter the lymph vessels.

e 4/4 marks awarded This is a detailed answer gaining full marks. The role of bile is described correctly, the vocabulary is appropriate and the digestion described in detail. Micelles are included. The absorption is also detailed, including the role of chylomicrons.

(ii) The person releases less bile as they don't have a gall bladder to store it in. Therefore, there is less bile to emulsify lipids. The lipase enzymes can only digest lipids ✓ on the outside of the fatty droplets, so lipid/fat digestion is less complete ✓.

e 2/2 marks awarded This answer is excellent and gains full marks. Student A has described the role of bile accurately and made it clear that although enzymes are still there, they are less effective without the emulsifying action of bile.

Student B

(a) Breaking down the food into smaller components.

e 0/2 marks awarded No marks are awarded here. The student doesn't say why this is necessary (for absorption) and 'breaking down' could be mechanical — it needs to refer to larger molecules being split into smaller ones at least.

(b) (i) They are broken down by enzymes secreted by the pancreas. Bile also breaks down lipids into fatty acids and glycerol. These diffuse into the epithelial cells of the small intestine.

e 0/4 marks awarded This answer is too vague. Student B has used the term 'broken down' rather than 'digested' or 'hydrolysed'. The exact enzyme (lipase) is not named. Then the student implies that bile digests lipids by saying that it 'breaks lipids down into fatty acids and glycerol'. This is clearly wrong. The way the products of digestion are absorbed is also too brief and lacks detail. This gains no marks.

(ii) This is because they don't produce bile any more so the fats don't get broken down properly.

e **0/2 marks awarded** The student is wrong in saying bile is not produced any more, but this doesn't lose the marks. The answer is too vague because the role of bile in fat digestion is not explained. Therefore, no marks are awarded.

Question 4

Two identical plants were taken. One had a ring of phloem removed below its bottom leaf and the other was left intact. A solution of phosphate ions containing radioactive phosphorus (^{32}P) was supplied to the roots of each plant. Both plants were grown in an atmosphere containing radioactive carbon dioxide ($^{14}CO_2$).

Table 1 shows the concentrations of ^{32}P and ^{14}C in the plants after 24 hours.

	Concentration/arbitrary units	
	Unringed plant	Ringed plant
^{32}P in roots	100	100
^{32}P in leaves	2.2	1.9
^{14}C in roots	0.9	0
^{14}C in leaves	0.1	0.99

Table 1

(a) **Phosphate ions are transported through the plant mainly in the xylem vessels. Explain the evidence from Table 1 to support this statement.** (2 marks)

(b) **Use evidence from Table 1 to explain how organic compounds are transported through a plant.** (3 marks)

In another investigation, a cooling jacket was placed round the petiole of a leaf, as shown in Figure 4. This resulted in organic materials being transported out of the leaf much more slowly.

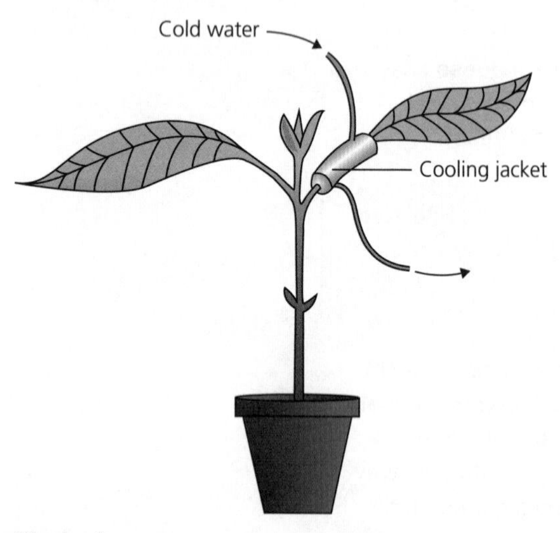

Figure 4

(c) **(i)** **Explain why the cooling jacket reduces the rate at which organic materials are transported out of the leaf.** (3 marks)

(ii) **Suggest a suitable control for this investigation, explaining why it is needed.** (2 marks)

ⓔ This is another question where you need to apply your knowledge and understanding to new situations. Notice that in (a) and (b) you are asked to use evidence from Table 1, so you will not get any credit unless you follow this instruction. Part (c) (i) is also application, while (c) (ii) tests your understanding of experimental design.

Student A

(a) The concentration of radioactive phosphorus in the leaves of both plants is similar, although it is a little lower in the ringed plant ✓. This must have travelled to the leaves in the xylem and not the phloem as the phloem is removed in a ringed plant ✓.

ⓔ **2/2 marks awarded** This answer gains both marks as student A has spotted the relevant information from the table and explained how the evidence supports the statement.

(b) The radioactive carbon remains in the leaves in the ringed plant ✓ and cannot travel to the roots, although it does travel to the roots in the unringed plant ✓. The only difference between the plants is that ringing removes the phloem so it must be the phloem that transports organic compounds ✓.

ⓔ **3/3 marks awarded** This answer gains all 3 marks as the student identifies the phloem as the tissue that transports organic compounds. The student has extracted the relevant information from the table and explained how this is relevant.

(c) (i) This slows down metabolism, which is controlled by enzymes ✓. Therefore, respiration is slower, which produces ATP for active transport ✓. This means that less ATP is available to transport organic materials into the phloem ✓.

ⓔ **3/3 marks awarded** This answer gains all 3 marks. Student A understands that cooling slows down the activity of enzymes in respiration, so less ATP will be made. In addition, the student also relates this to the loading of organic compounds into the phloem.

(ii) The other plant would be treated exactly the same, but the cooling jacket would not be turned on ✓. This would show it is the colder temperature that has the effect and not just putting a jacket around the petiole ✓.

ⓔ **2/2 marks awarded** This answer gets full marks as the correct control, and the reason for it, is given.

Student B

(a) In the ringed plant, slightly less radioactive phosphorus reaches the leaves than in the unringed plant ✓. The xylem is still intact in the ringed plant, so the phosphorus in the ringed plant must have reached the leaves via the xylem and not the phloem ✓.

e **2/2 marks awarded** This answer also gains both marks as student B has spotted the relevant information from the table and explained how the evidence supports the statement.

(b) They travel in the phloem ✓. Organic compounds contain carbon. The radioactive carbon cannot get to the roots in the ringed plant ✓, showing that organic compounds travel in the phloem but the radioactive carbon remains in the leaves ✓.

e **3/3 marks awarded** This answer gains all 3 marks as the student B has identified the relevant information from the table. The student clearly explains how the absence of radioactive carbon in the roots, and the high concentration of radioactive carbon in the leaves, shows that organic compounds must travel in the phloem.

(c) (i) The cooling jacket slows down the rate of enzyme activity ✓, so respiration will be slower. This means less ATP is available ✓.

e **2/3 marks awarded** This answer is awarded 2 marks, for relating the cooling to enzymes and respiration, and therefore knowing that less ATP is available. However, the student does not get a third mark as there is no reference to what the ATP is used for.

(ii) Repeat this with a cooling jacket that is not switched on ✓. This is to show that it is the cooler temperature that affects the loading of organic compounds into the phloem and not anything else ✓.

e **2/2 marks awarded** This answer also gets both marks for identifying the correct control and why it is needed.

Question 5

(a) What is a gene? (1 mark)

(b) The DNA code is said to be universal, non-overlapping and degenerate. Complete the table to explain what each of these words means. (3 marks)

Term	Definition
Universal	
Non-overlapping	
Degenerate	

Figure 5 shows protein synthesis taking place in a ribosome.

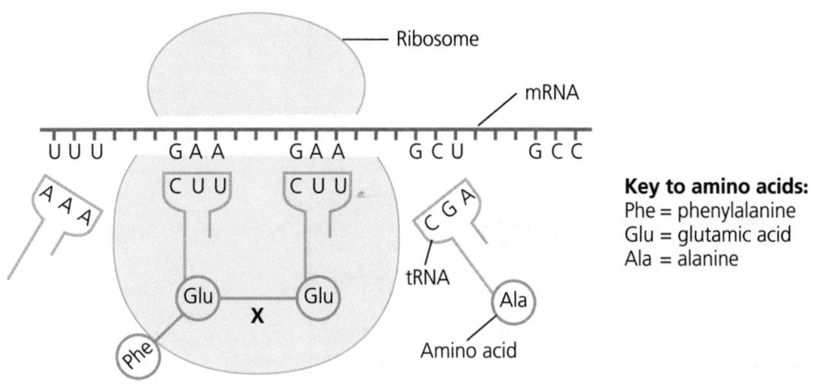

Key to amino acids:
Phe = phenylalanine
Glu = glutamic acid
Ala = alanine

Figure 5

(c) (i) Name the bond labelled X. (1 mark)

(ii) Use examples from the diagram to explain the terms 'codon' and 'anticodon'. (3 marks)

ⓔ This question tests your knowledge of DNA and protein synthesis. This should be fairly straightforward if you understand the process. Note that in (c) (ii) you need to use information from the diagram in your answer.

Student A

(a) A base sequence of DNA at a specific locus on a chromosome that codes for a polypeptide ✓ or a functional RNA molecule.

ⓔ **1/1 mark awarded** This answer gets 1 mark as only 1 mark is available. However, it is a full answer.

(b)

Term	Definition
Universal	The DNA code is the same in every living thing ✓
Non-overlapping	Each base is part of only one codon ✓
Degenerate	There are more codons than amino acids, so some amino acids are coded for by more than one codon ✓

ⓔ **3/3 marks awarded** This answer gains all 3 marks, as each term is well explained.

(c) (i) Peptide bond ✓.

ⓔ **1/1 mark awarded** This answer gains the mark. It is a straightforward question.

(ii) GAA on the messenger RNA is a codon ✓ and CUU on the tRNA is an anticodon ✓. They are complementary to each other ✓.

ⓔ **3/3 marks awarded** This answer refers to the diagram and infers that each is a triplet of bases. The third mark, for saying that they are complementary, is also awarded.

Student B

(a) A piece of DNA that codes for a protein ✓.

e **1/1 mark awarded** This answer gets the mark (just), but it would be better to say it codes for a polypeptide or functional RNA.

(b)

Term	Definition
Universal	It is the same in all living organisms ✓
Non-overlapping	Each code doesn't overlap with the next code
Degenerate	Some amino acids have several triplets and others have only one ✓

e **2/3 marks awarded** This answer gains 2 marks, one for 'universal' and one for 'degenerate'. However, 'non-overlapping' is not described well. It is generally not a good idea to reuse the terms in the question — by saying that each code doesn't overlap, student B doesn't show that the concept of overlapping is understood. If the student had written 'the code is read three bases at a time, then the next three, and so on' this would show that the term was clearly understood.

(c) **(i)** Peptide bond ✓.

e **1/1 mark awarded** This answer gains the mark.

(ii) The codon is on messenger RNA and the anticodon is on the tRNA.

e **0/3 marks awarded** This answer doesn't tell us that the codon and anticodon are triplets of bases and there is no reference to the diagram, so this gets no marks.

Question 6

Figure 6 shows how blood pressure changes as blood flows through the circulatory system.

(a) **(i)** **Explain why the pressure increases and decreases regularly as blood flows through the arteries.** (2 marks)

(ii) **Explain why the pressure drops rapidly as blood flows through the arterioles and capillaries.** (2 marks)

(b) **(i)** **Describe how tissue fluid is formed and returned to the blood.** (5 marks)

(ii) **Some people have a liver disease that means the liver makes very little albumin. Albumin is an important plasma protein. People with this disease have oedema (swelling) in their legs. Explain why.** (3 marks)

e Part (a) tests your understanding of blood circulation, whereas (b) (i) is straightforward recall. This is followed by (b) (ii), which applies your understanding to a new situation that you are probably unfamiliar with.

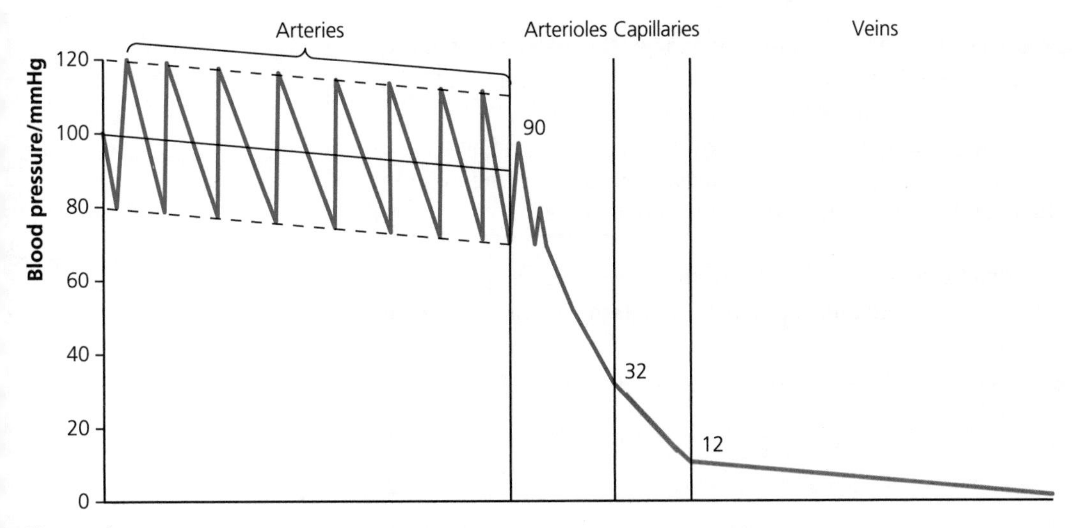

Figure 6

Student A

(a) (i) The peak is when ventricular systole occurs ✓ and the decrease in pressure is diastole ✓.

e 2/2 marks awarded This answer gets both marks. It explains the peaks and the troughs in the graph, and also relates it to the cardiac cycle.

(ii) There are many arterioles and capillaries and, although they are small in diameter, there are so many of them that the total cross-sectional area is greater than the arteries so the pressure drops ✓. Pressure also falls as a result of friction ✓.

e 2/2 marks awarded This answer gets both marks. The fall in pressure is related to the cross-sectional area and also to friction.

(b) (i) At the arteriole end of the capillary, water and small molecules are forced out of the capillary as a result of the pressure from the beating of the heart ✓, forming tissue fluid. Large proteins remain in the capillary ✓. This lowers the water potential at the arteriole end of the capillary ✓, so that water from the tissue is reabsorbed by osmosis, carrying solutes with it ✓. The rest of the tissue fluid drains into the lymph vessels ✓ and eventually returns to the blood.

e 5/5 marks awarded This answer gets all 5 marks. The formation of tissue fluid under pressure from the heart beating is clear, and the return of tissue fluid to the lymph vessels and blood system is also clear. The role of plasma proteins is included.

(ii) In a normal person, the plasma proteins lower the water potential of the blood at the venule end ✓, as they are too large to enter the tissue fluid. If the person does not have enough plasma protein, there is a smaller water potential gradient between the tissue fluid and the blood ✓. Therefore, less water is reabsorbed by osmosis, meaning that tissue fluid builds up in the tissues ✓, causing oedema.

e 3/3 marks awarded Student A is clear about the role of plasma proteins in returning tissue fluid to the blood and explains how oedema forms clearly. This response gets full marks.

Student B

(a) (i) This is the pressure of the heart beating ✓.

e 1/2 marks awarded Student B relates the pressure to the cardiac cycle, so gets 1 mark. However, there is not enough description for the second mark.

(ii) This is because the arterioles and capillaries are narrower and there is more friction ✓.

e 1/2 marks awarded 1 mark is awarded for friction ✓.

(b) (i) Tissue fluid is forced out of the capillary at the artery end. The tissue fluid carries nutrients and oxygen to the cells. Some of the tissue fluid returns to the blood via the lymph vessels ✓. The rest of the tissue fluid passes into the capillary at the vein end by osmosis ✓.

e 2/5 marks awarded This answer gets only 2 marks. It is not clear that it is pressure from the heart forcing the tissue fluid out of the capillary, nor that it is formed from blood plasma. The fact that tissue fluid returns to the blood via lymph vessels is worth 1 mark. The student is wrong to say tissue fluid returns to the capillary by osmosis — it is *water* that enters by osmosis — and the role of plasma proteins in creating the water potential gradient is not mentioned. However, 1 mark is given for osmosis.

(ii) This means that more tissue fluid is made and it cannot all return to the blood. This causes oedema ✓.

e 1/3 marks awarded Student B doesn't understand the role of plasma proteins in returning tissue fluid to the blood. This response is a guess, which is wrong. However, 1 mark is given for explaining that oedema is a build-up of tissue fluid.

Question 7

Figure 7 shows the pressure changes in the aorta, left ventricle and left atrium during the cardiac cycle.

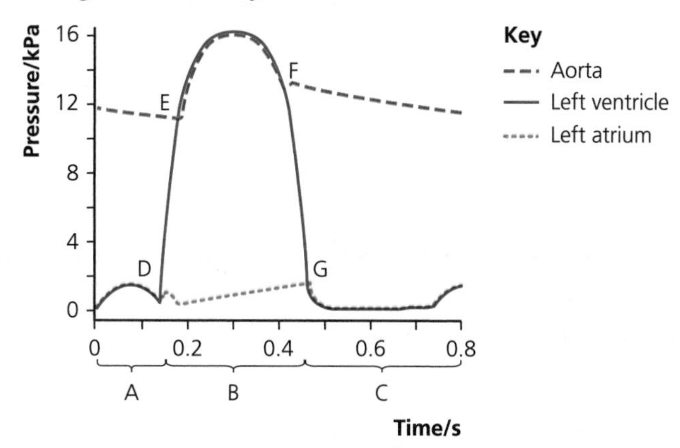

Figure 7

(a) **Complete the table with the correct letter to show the time when each of the following events occur.** (4 marks)

Letter	Event
	The point where the semilunar valves open
	A time when the ventricle is contracting
	The point where the atrioventricular valve closes
	The point where the atrioventricular valve opens

(b) **This person's cardiac output is 4900 $cm^3 min^{-1}$. Calculate this person's stroke volume.** (3 marks)

ⓔ This question checks your understanding of the cardiac cycle. It should be straightforward if you understand the topic. The calculation in (b) should be easy if you remember the formula involved.

Student A

(a)

Letter	Event
E	The point where the semilunar valves open ✓
B	A time when the ventricle is contracting ✓
D	The point where the atrioventricular valve closes ✓
G	The point where the atrioventricular valve opens ✓

ⓔ **4/4 marks awarded** This student understands the principles and therefore gains full marks.

(b) Cardiac output = stroke volume × heart rate

Therefore:

$$\text{Stroke volume} = \frac{\text{cardiac output}}{\text{heart rate}}$$

$$\text{Heart rate} = \frac{60}{0.7}$$

$= 86$ beats per min ✓

$$\text{Stroke volume} = \frac{4900}{86}$$

$= 57\,cm^3$ ✓✓

e 3/3 marks awarded This answer gets full marks. The working is shown, including a correct calculation of heart rate, so even if the final answer had been wrong student A would have gained at least 1 mark.

Student B

(a)

Letter	Event
F	The point where the semilunar valves open ✗
B	A time when the ventricle is contracting ✓
G	The point where the atrioventricular valve closes ✗
D	The point where the atrioventricular valve opens ✗

e 1/4 marks awarded This answer gets only 1 mark, for a time when the ventricle is contracting. Student B has tried to memorise the graph rather than understand it. The semilunar valves open when the pressure in the ventricle is higher than the pressure in the aorta, so the valve opens at E (F is the time when the valve closes). The atrioventricular valve is the wrong way round too. It opens when the pressure in the atrium rises above that of the ventricle (at G) and closes when the pressure in the ventricle is greater than in the atrium (at D).

(b) $49\,cm^3$ ✗

e 0/3 marks awarded Student B has given an incorrect answer. In addition, no working is shown. 1 mark is available for calculating the heart rate correctly, but the student has not given any figure for this.

Question 8

Scientists have recently found a new species of Australian humpback dolphin, *Sousa sahulensis*. To decide whether this was a new species, scientists used a lot of information about humpback dolphins, including a study of their DNA.

The Australian humpback dolphin differs in size, number of teeth and vertebrae, and in colouration from the other three humpback species.

(a) Complete the table to show the classification of the new species of humpback dolphin. (2 marks)

Taxon	Australian humpback dolphin
Domain	Eukarya
Kingdom	Animalia
	Chordata
	Mammalia
	Cetacea
	Delphinidae
Genus	
Species	

(b) (i) How would DNA evidence be used to show that this is a new species? (2 marks)

(ii) What would the scientists need to do to prove that this is a new species? (2 marks)

(c) The scientists believe that natural selection has resulted in the evolution of this different population of humpback dolphins. Explain how. (4 marks)

ⓔ The table of classification is straightforward recall. You are not expected to know the genus and species of this animal, but to use the information in the stem of the question. Parts (b) and (c) test your understanding in this specific context.

Student A

(a)

Taxon	Australian humpback dolphin
Domain	Eukarya
Kingdom	Animalia
Phylum	Chordata
Class	Mammalia
Order	Cetacea
Family	Delphinidae
Genus	Sousa
Species ✓	sahulensis ✓

ⓔ **2/2 marks awarded** This answer gets both marks. Both columns are correct, and student A has even remembered to underline the genus and species names.

(b) (i) They would extract DNA from the known species of dolphins and this new kind. They would compare the base sequences ✓ and see how many differences there are ✓. If there are many differences, it is likely to be a new species.

ⓔ **2/2 marks awarded** This gets both marks as the base sequence is found and a method of comparing them is also mentioned.

(ii) They would interbreed them with the other kinds of dolphins ✓ to find out whether they can produce fertile offspring ✓. If they produce fertile offspring with the other dolphins, they are not a new species.

ⓔ **2/2 marks awarded** This is a full and correct answer, scoring both marks.

(c) The original dolphin population showed variation in size, shape and colour ✓. In one area, the larger dolphins were at an advantage. This meant larger dolphins were more likely to survive ✓ and so had more offspring ✓. Their offspring inherited the allele for large size ✓. In another area, smaller dolphins with a different colour were at an advantage, so these were the dolphins that survived, reproduced and passed on their alleles to their offspring.

ⓔ **4/4 marks awarded** This is a good answer with technical vocabulary. Student A mentions that more successful dolphins survive, reproduce and pass on their alleles. This makes it a full and detailed answer and gets full marks.

Student B

(a)

Taxon	Australian humpback dolphin
Domain	Eukarya
Kingdom	Animalia
Phylum	Chordata
Class	Mammalia
Order	Cetacea
Family	Delphinidae
Genus	
Species ✓	

ⓔ **1/2 marks awarded** Student B gets 1 mark for the first column, but nothing for the second column. The student didn't realise that the genus and species are given in the question.

(b) (i) They would find the base sequence ✓ of the DNA of this dolphin and compare it with the base sequence of the DNA from other dolphins.

ⓔ **1/2 marks awarded** This gets 1 mark for finding the base sequence of the DNA, but it is not clear how comparing it will help. Student B needs to mention finding the number of differences, for example.

(ii) They would interbreed ✓ them to see if they can produce offspring.

e **1/2 marks awarded** This gets 1 mark for interbreeding, but simply producing offspring doesn't show whether they are the same or a different species. The offspring must be fertile if they are the same species.

(c) The original population of dolphin were a little different in size and shape ✓. In one area it was an advantage to be larger or to have different coloration. The dolphin that were larger were selected for ✓ and so more of the dolphins became larger.

e **2/4 marks awarded** This is a vague answer that fails to use technical vocabulary. Student B does mention that the original population showed variation, although using the word 'variation' might be better. There is also the idea of selection, so this answer only gets 2 marks.

Question 9

(a) Describe how water passes into a plant root and enters the xylem. (5 marks)

The creosote bush is a xerophytic desert plant. It has many flexible stems that spread out from its base. It produces biochemical compounds that taste and smell unpleasant. It has small leaves with stomata that close by day and open at night. It has a double-root system that extends a long way sideways and also deep down.

(b) Select two features of the creosote bush from the information above that are adaptations to reduce water loss. For each feature, explain how it helps to reduce water loss. (4 marks)

e Part (a) is recall of what you have learned and it should be easy to gain full marks. Note that (b) asks for features mentioned in the question, so don't just give general xerophytic adaptations that you may have learned.

Student A

(a) The water in the soil has a higher water potential than the root hair cell, so water enters the root hair cell by osmosis ✓. Some of the water crosses the cortex by the symplast pathway, where water moves from cell to cell by osmosis ✓ and via the plasmodesmata. Some of the water moves via the apoplast pathway, which is where it travels through the cell walls and extracellular spaces ✓. When it reaches the endodermis, the water can only move via the symplast, as the apoplast is blocked by the Casparian strip ✓. Water then enters the xylem down a water potential gradient ✓, created by ions being pumped into the xylem.

e **5/5 marks awarded** This is an excellent answer gaining all 5 marks. Every stage is well described and technical vocabulary is used appropriately.

(b) Feature 1 — small leaves ✓

Explanation — reduces the surface area from which water can evaporate ✓

Feature 2 — stomata that open at night and close by day ✓

Explanation — it is cooler at night, so less water is lost by opening the stomata at night instead of by day ✓

e **4/4 marks awarded** This gets all 4 marks. Two suitable features are chosen and both are well explained.

Student B

(a) Water enters the root hair cells by osmosis. The water travels through the root cortex by the symplast and apoplast pathways ✓. At the endodermis, all the water is forced to move by the symplast pathway. Ions are pumped into the xylem, creating a water potential gradient so water enters the xylem by osmosis ✓.

e **2/5 marks awarded** This response gets only 2 marks. Although student B has mentioned water entering the root hair cell by osmosis, there is no reference to the water potential gradient. The symplast and apoplast pathways are mentioned, but the student does not say which is which and how they are different. The endodermis is mentioned, but not how it forces water to move by the symplast. However, the way water enters the xylem is well described.

(b) Feature 1 — biochemical compounds that taste and smell unpleasant

Explanation — stops the plant from being eaten so that it can survive better

Feature 2 — double-root system ✓

Explanation — can absorb water from a long way down and also from a long way sideways if there is some rain ✓

e **2/4 marks awarded** This gets 2 marks. The first feature chosen is not an adaptation to reduce water loss, so it gets no marks. The second adaptation is correctly chosen and well explained.

Test paper 2

This paper has 91 marks available and should take 2 hours to complete. It is similar in style to A-level paper 1. There is an overlap in the content covered in this paper and the AS paper, so this paper can be used for extra practice if you are studying AS.

Question 1

Figure 1 shows the oxyhaemoglobin dissociation curve for two kinds of fish.

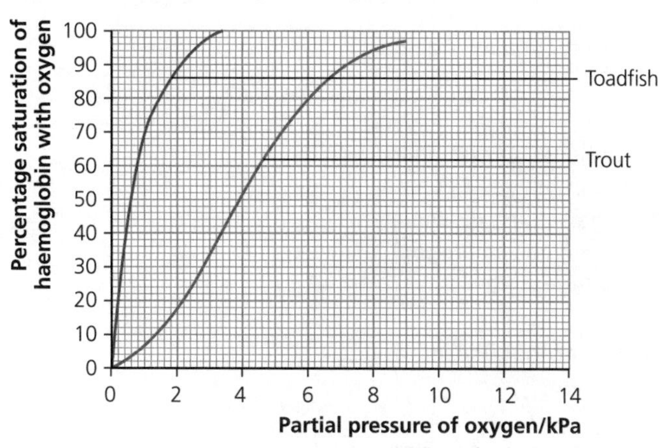

Figure 1

(a) (i) Assume the trout's haemoglobin is 100% saturated in the gills. The partial pressure of oxygen in the fish's liver is 4 kPa. How much oxygen does the haemoglobin release in the fish's liver? (1 mark)

(ii) The trout lives in faster-flowing water containing more oxygen than the toadfish. Explain how the haemoglobin of each fish adapts it to its environment. (4 marks)

(b) Table 1 shows some data about the gills of some kinds of fish.

Type of fish	Activity level of fish	Number of secondary lamellae per mm of filaments	Gill area/$mm^2 g^{-1}$
Toadfish		11.00	151
Mackerel		29.40	1040
Mullet	Active	28.50	1010
Tautog	Intermediate	21.58	461
Fluke	Slow	17.30	247
Sea robin	Intermediate	20.00	432

Table 1

(i) Complete the table to show the activity levels of toadfish and mackerel. Give a reason for your answers. (2 marks)

(ii) Explain why the gill area is given per gram of fish. (2 marks)

(c) There is a counter-current flow of water and blood across the gill lamellae in a fish. Explain the advantage of this. (2 marks)

e Part (a) (i) asks for a simple figure that you can read off the graph. Part (a) (ii) asks you to apply your knowledge of oxyhaemoglobin dissociation curves to these specific examples. Part (b) tests your ability to interpret data and relate them to your understanding of gas exchange in fish. Part (c) is straightforward and simply tests recall.

Student A

(a) (i) 50% ✓

e **1/1 mark awarded** This is an easy question and the correct answer gains 1 mark.

(ii) The toadfish has haemoglobin with a higher affinity for oxygen than trout haemoglobin ✓. This means that the haemoglobin can become fully saturated at lower partial pressures of oxygen ✓ and dissociates readily at lower partial pressures of oxygen ✓. Trout haemoglobin dissociates readily, which enables a fast respiration rate ✓.

e **4/4 marks awarded** Full marks for a fully correct answer that uses scientific terminology well.

(b) (i)

Type of fish	Activity level of fish	Number of secondary lamellae per mm of filaments	Gill area/$mm^2\,g^{-1}$
Toadfish	Slow	11.00	151
Mackerel	Active	29.40	1040
Mullet	Active	28.50	1010
Tautog	Intermediate	21.58	461
Fluke	Slow	17.30	247
Sea robin	Intermediate	20.00	432

The toadfish has a lower surface area of gills than the fluke, which is also slow ✓. Mackerel have a higher surface area than mullet, which is active ✓.

e **2/2 marks awarded** This is correct and gets both marks.

(ii) This is because the fish are different sizes ✓ so it allows you to compare ✓ them.

e **2/2 marks awarded** This gets both marks.

(c) It maintains a concentration gradient ✓ all the way along the lamellae ✓.

e **2/2 marks awarded** This gets both marks for a full explanation.

Student B

(a) (i) 50% ✓

ⓔ **1/1 mark awarded** This is an easy question and the correct answer gains 1 mark.

(ii) Toadfish haemoglobin is fully saturated with oxygen at a lower oxygen concentration than trout haemoglobin ✓. Trout haemoglobin releases its oxygen readily, which enables the trout to respire rapidly ✓.

ⓔ **2/4 marks awarded** Student B has not used technical words such as 'affinity', 'dissociates' or 'partial pressure'. They have omitted toadfish haemoglobin dissociating at low partial pressures of oxygen and have not said which haemoglobin has the higher affinity for oxygen.

(b) (i)

Type of fish	Activity level of fish	Number of secondary lamellae per mm of filaments	Gill area/$mm^2 g^{-1}$
Toadfish	Slow	11.00	151
Mackerel	Active	29.40	1040
Mullet	Active	28.50	1010
Tautog	Intermediate	21.58	461
Fluke	Slow	17.30	247
Sea robin	Intermediate	20.00	432

The toadfish has the smallest surface area of any fish on the table ✓ and mackerel has the largest ✓.

ⓔ **2/2 marks awarded** This is correct and gets both marks.

(ii) Because some fish are larger than others ✓.

ⓔ **1/2 marks awarded** This answer gets 1 mark but not the second mark because the idea of comparing different fish is not mentioned.

(c) The water and the blood move in opposite directions, making gas exchange more efficient.

ⓔ **0/2 marks awarded** This gets no marks because student B is describing counter-current flow rather than explaining it, as the question asks. Just saying that gas exchange is more efficient is not an explanation — a reason *why* is needed.

Question 2

Figure 2 shows the rate of transpiration and water absorption by a plant over a 24-hour period in the summer.

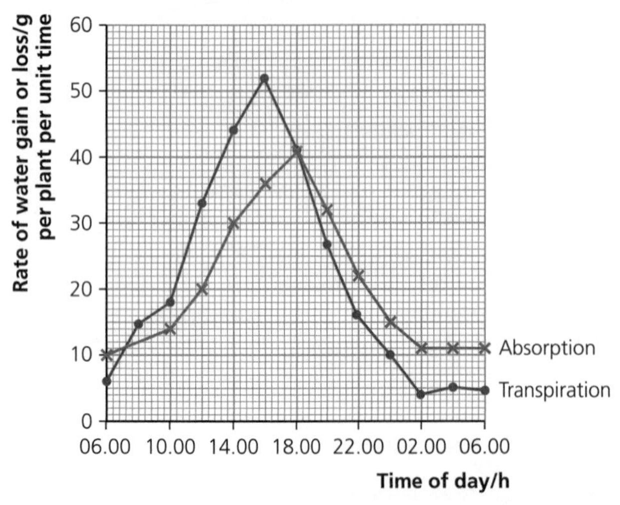

Figure 2

(a) **At what time of day is the difference between the rate of transpiration and the rate of absorption at its greatest?** (1 mark)

(b) **Explain the rate of transpiration:**

(i) **between 12.00 and 16.00 hours** (2 marks)

(ii) **between 02.00 and 06.00 hours** (1 mark)

(c) **Describe how the structure of the endodermis affects water transport across the root of a plant.** (3 marks)

ⓔ Parts (a) and (b) assess your ability to apply what you have learned to a specific graph. This is followed by simple recall in part (c).

Student A

(a) From 15.30 to 16.00 hours ✓.

ⓔ **1/1 mark awarded** The correct answer gains 1 mark.

(b) **(i)** The temperature increases ✓ so evaporation from the stomata, which are open, increases with temperature ✓.

ⓔ **2/2 marks awarded** The correct answer gains both marks.

(ii) The stomata are closed because the plant is not photosynthesising so transpiration is much lower ✓.

ⓔ **1/1 mark awarded** The correct answer gains 1 mark.

(c) The endodermis has a Casparian strip ✓ made of suberin (a waxy substance) that blocks the apoplast pathway ✓, forcing all the water to pass via the symplast pathway ✓.

ⓔ **3/3 marks awarded** This gains all 3 marks for a full and clear answer.

Student B

(a) Between 14.00 and 16.00 hours.

ⓔ **0/1 mark awarded** This answer is not precise enough for 1 mark.

(b) (i) The stomata are open so transpiration increases ✓ as temperature increases ✓.

ⓔ **2/2 marks awarded** The correct answer gains both marks.

(ii) The stomata close so transpiration is reduced ✓.

ⓔ **1/1 mark awarded** The correct answer gains 1 mark.

(c) It stops water moving through the apoplast ✓ because it has a waxy substance in its cell wall ✓.

ⓔ **2/3 marks awarded** This gets 2 marks — one for the apoplast being blocked and the other for mentioning the waxy substance in the cell wall. However, it does not refer to all the water travelling via the symplast pathway from this point on.

Question 3

Table 2 gives some information about the composition of the walls of different kinds of blood vessels.

	Composition of blood vessel wall/%			
	Endothelium	Smooth muscle	Elastic tissue	Connective tissue
Artery	5	25	40	30
Arteriole	10	60	10	20
Capillary	95	–	–	5
Venule	20	20	–	60

Table 2

(a) (i) Arteries contain a high proportion of elastic tissue. Explain the advantage of this. (2 marks)

(ii) Arterioles contain a high proportion of smooth muscle. Explain the advantage of this. (2 marks)

(iii) Capillaries contain no smooth muscle or elastic tissue. Explain the advantage of this. (2 marks)

Figure 3 summarises how tissue fluid is formed and returned to the capillaries or lymph vessels.

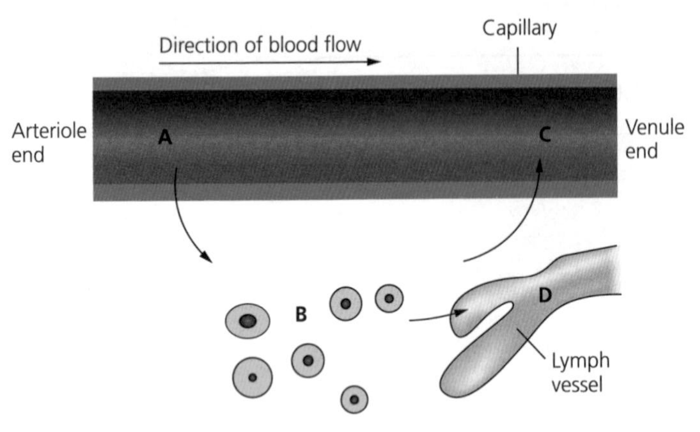

Figure 3

(b) **Choose which letter from Figure 3 — A, B, C or D — indicates:**

(i) **a point with the lowest water potential**

(ii) **a point where the oxygen concentration is highest** (2 marks)

(c) **What causes tissue fluid to form at the arterial end of a capillary?** (1 mark)

ⓔ This question tests both your knowledge and understanding of blood vessels and tissue fluid formation, and your ability to apply your understanding to a new situation.

Student A

(a) **(i)** This allows elastic recoil ✓, which keeps the blood under pressure ✓.

ⓔ **2/2 marks awarded** This is an excellent answer that gets both marks. Student A makes it clear by the use of the term 'elastic recoil' that they know the role of the elastic fibres.

(ii) The muscle can contract (causing vasoconstriction) or dilate (causing vasodilation) ✓. This means that the blood flow to different parts of the body can be modified ✓.

ⓔ **2/2 marks awarded** This is a good answer as it explains how the muscles can alter blood flow to different parts of the body. This answer gets the full 2 marks.

(iii) This means the wall of the capillary is very thin ✓ to allow a short diffusion pathway for substances to be exchanged with body tissues ✓.

ⓔ **2/2 marks awarded** This is a good answer. It relates the structure to the role of capillaries in exchanging substances between the tissues and the blood. It also relates this to a short diffusion pathway, gaining both available marks.

(b) (i) C ✓

(ii) A ✓

e **2/2 marks awarded** This is a good answer that gets both marks.

(c) The pumping of the heart causes tissue fluid ✓ to form by ultrafiltration from blood plasma.

e **1/1 mark awarded** This is a good answer that gets 1 mark. Student A has even added extra detail, although no extra marks are awarded for this.

Student B

(a) (i) This helps to push the blood along so it is under pressure ✗.

e **0/2 marks awarded** This gets no marks because student B implies that the walls of the artery pump the blood — which, of course, is wrong, as only the heart can pump the blood. Although the student says that this keeps the blood under pressure, it is in the wrong context.

(ii) This allows blood flow to different organs to be controlled ✓.

e **1/2 marks awarded** This scores 1 mark, but there is no explanation of how it controls blood flow and so no second mark is gained.

(iii) They carry blood at low pressure and don't need to control blood flow.

e **0/2 marks awarded** This is not worth any marks. The advantage of the thin walls has not been explained. The answer must be related to the role of the capillaries.

(b) (i) C ✓

(ii) A ✓

e **2/2 marks awarded** This is a correct answer that gets both marks.

(c) The pressure of the blood forces water and dissolved substances from the plasma to be forced out ✓.

e **1/1 mark awarded** This is a good answer and gets 1 mark, although it doesn't have the extra detail that student A included.

Question 4

(a) What is meant by the following terms?

(i) triplet

(ii) intron (2 marks)

Enzyme A and enzyme B both catalyse the same reaction. However, enzyme A is more efficient than enzyme B. Both enzymes have the same number of amino acids in their polypeptide chain. The enzymes have exactly the same structure except for a small part of the chain. The amino acid sequence of this part of enzyme A is shown in Table 3, together with the corresponding base sequence.

Amino acid sequence for enzyme A	Val	His	Leu	Asn	Lys	Leu	Ser	Ala
DNA sequence for enzyme A	CAG	GTA	GAA	TTA	TTT	GAA	AGT	CGA
mRNA sequence for this part of enzyme A								

Table 3

(b) Complete Table 3 to show the appropriate mRNA sequence for this part of enzyme A. (1 mark)

It was found that enzyme B was the result of two substitution mutations in the DNA coding for enzyme A. The amino acid sequence of the corresponding part of enzyme B is:

Val Leu Leu Asn Asn Leu Ser Ala

Table 4 shows part of the genetic code.

Amino acid	Codons that can be used
Ala	GCU
Asn	AAU
His	CAU or CAC
Leu	CUU or UUA
Lys	AAA
Met	AUG
Pro	CCA
Ser	AGU or UCA
Val	GUC

Table 4

(c) **(i)** Use Table 4 to work out, as accurately as you can, the sequence of bases in the mRNA coding for enzyme B. (2 marks)

(ii) Circle the bases (nucleotides) substituted in the two mutations. (1 mark)

(iii) Suggest a reason why enzyme B is not as efficient as enzyme A. (2 marks)

e Part (a) is simple recall, but parts (b) and (c) explore your understanding in a new context. This does require a little thought, but it should be accessible to students aiming for higher grades.

Student A

(a) (i) Three bases in the DNA that code for an amino acid ✓.

(ii) A non-coding region of the DNA ✓.

e 2/2 marks awarded This is a good answer that gets both marks.

(b)

Amino acid sequence for enzyme A	Val	His	Leu	Asn	Lys	Leu	Ser	Ala
DNA sequence for enzyme A	CAG	GTA	GAA	TTA	TTT	GAA	AGT	CGA
mRNA sequence for this part of enzyme A	GUC	CAU	CUU	AAU	AAA	CUU	UCA	GCU ✓

e 1/1 mark awarded This gets 1 mark for a correct answer.

(c) (i) GUC CUU CUU AAU AAU CUU UCA GCU ✓✓

e 2/2 marks awarded This gets both marks as it is the same as the mRNA coding for the relevant part of enzyme A, but with two changes to fit the amino acid sequence given for enzyme B.

(ii) GUC C(U)U CUU AAU AA(U) CUU UCA GCU ✓

e 1/1 mark awarded This gets the mark for a fully correct answer.

(iii) This sequence is in the active site ✓, so the active site is a slightly different shape meaning the substrate does not fit ✓ in it as well as enzyme A.

e 2/2 marks awarded This gets both marks, 1 for referring to the active site and the second for explaining why a changed shape may not be as efficient.

Student B

(a) (i) Three bases in DNA coding for an amino acid ✓.

(ii) A non-coding region of the DNA. This is spliced out of mRNA before translation ✓.

e 2/2 marks awarded This is a good answer that gains both marks.

(b)

Amino acid sequence for enzyme A	Val	His	Leu	Asn	Lys	Leu	Ser	Ala
DNA sequence for enzyme A	CAG	GTA	GAA	TTA	TTT	GAA	AGT	CGA
mRNA sequence for this part of enzyme A	GAC	CAT	CTT	AAT	AAA	CTT	TCA	GCT ✗

e 0/1 mark awarded This gets no marks, unfortunately, as student B has found a complementary sequence but has forgotten that, in RNA, thymine is replaced by uracil.

(c) (i) GUC UUA AAU AAU UUA UCA GCU ✓

e 1/2 marks awarded This gets 1 mark for giving a code that would code for the amino acids required. However, this is not based on a mutated version of the gene for enzyme A so the second mark is not given.

(ii) GUC UUA AAU AAU (UUA) (UCA) GCU ✗

e 0/1 mark awarded This gets no marks as a substitution is not a change in all three bases in a triplet.

(iii) It is a different shape so it doesn't work as well.

e 0/2 marks awarded This gets no marks as there is no reference to the active site or the idea of shape and fit.

Question 5

A student wanted to investigate the effect of several antibiotics on the growth of a species of bacterium. She used aseptic technique to spread a culture of bacteria on the surface of an agar plate. She then placed four paper discs on the bacteria, each containing a different antibiotic. She incubated the plate for 24 hours. Figure 4 shows the results.

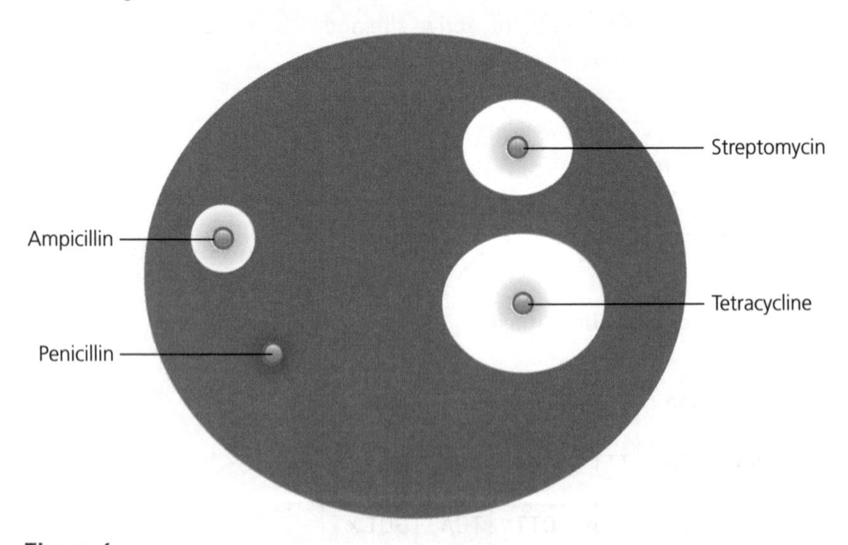

Figure 4

(a) **Give two examples of aseptic technique that the student might have used.** (2 marks)

(b) **There are clear zones around some of the paper discs. Explain how these have occurred.** (2 marks)

(c) **The student decided that, if these bacteria had been pathogenic bacteria from a person with a sore throat, tetracycline would be the best antibiotic to prescribe for the person. However, a doctor might choose a slightly less effective antibiotic instead. Suggest a reason why.** (1 mark)

(d) **Tetracycline binds to the mRNA codon and prevents the tRNA anticodon binding. Explain how this inhibits bacterial growth.** (3 marks)

(e) **The student decided that the bacteria she had used were resistant to penicillin. Describe how natural selection can lead to the spread of bacteria that are resistant to antibiotics such as penicillin.** (3 marks)

e Parts (a), (b) and (c) test your understanding of this practical technique. Part (d) tests your understanding in a new context, but (e) is recall again.

Student A

(a) Sterilising the forceps in a Bunsen flame before use ✓ and dipping the spreader in alcohol and then flaming it ✓.

e **2/2 marks awarded** This gets both marks for two good examples.

(b) Antibiotic has diffused out of the disc into the agar ✓, giving a clear zone where bacteria have been killed or inhibited ✓.

e **2/2 marks awarded** This gets both marks for a full, clear answer.

(c) Another antibiotic might have fewer side effects ✓.

e **1/1 mark awarded** This gets 1 mark. Note that any sensible suggestion will gain the mark.

(d) This means that proteins cannot be made ✓ as tRNA cannot bring amino acids to the ribosome ✓. These proteins may be for growth or enzymes ✓. Without these proteins, the bacteria will die.

e **3/3 marks awarded** Student A gains all 3 marks for explaining the effect on protein synthesis and the effect this will have.

(e) When penicillin is used, bacteria that are not resistant are killed ✓. By chance, one or two bacteria had a mutation, making them resistant to penicillin ✓. These bacteria survived, reproduced and passed on the resistance gene ✓, producing a whole population of bacteria that were resistant to penicillin.

e **3/3 marks awarded** This gains all 3 marks for a well-expressed answer.

Student B

(a) Cleaning the bench before use and wearing a lab coat.

e 0/2 marks awarded This gets no marks — something can be regarded as 'clean' without necessarily being free of bacteria. If the student had said 'wiping down the bench with bleach or alcohol', this would have gained credit. Wearing a lab coat is standard lab procedure and is not an extra measure taken in aseptic technique.

(b) This is where the antibiotic has killed the bacteria ✓.

e 1/2 marks awarded This gets only 1 mark, but it doesn't explain that the antibiotic has diffused into the agar.

(c) A different antibiotic might be a lot cheaper ✓ for the NHS.

e 1/1 mark awarded This gets 1 mark as it is a sensible suggestion.

(d) This stops protein synthesis so the bacteria cannot grow.

e 0/3 marks awarded This gets no marks as it is too vague. Student B needs to say how it stops protein synthesis and why stopping protein synthesis prevents bacterial growth.

(e) A mutation occurred that made a bacterium immune to penicillin. The other bacteria died but the immune bacteria survived ✓. This bacterium reproduced and passed on the gene for penicillin resistance to its offspring ✓.

e 2/3 marks awarded This would have gained all 3 marks, but the student has used the word 'immune' instead of 'resistant'. 'Immune' means that antibodies are made, but bacteria don't have an immune system. Therefore, one of the marks is not awarded and this gains only 2 marks.

Question 6

(a) A student was asked to find the species diversity index of plant species in a meadow. What measurements would he need to make? (2 marks)

(b) The student used a quadrat frame to sample the meadow. Describe how he should have done this. (3 marks)

In organic farming, chemical pesticides and artificial fertilisers are not used. A journalist examined many scientific investigations that compared biodiversity in organic farms and conventional farms. The results are shown in Figure 5. The numbers in the circles indicate the number of studies that showed there was no impact.

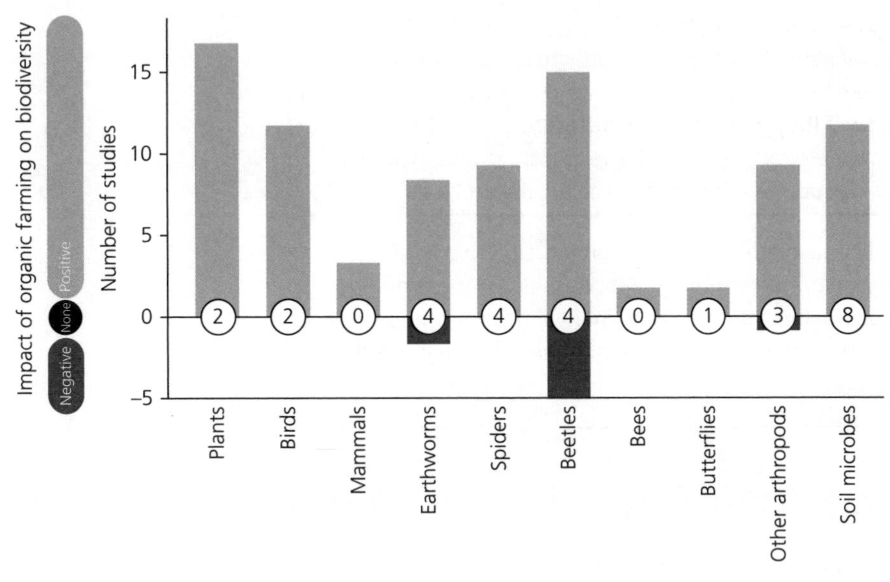

Figure 5

(c) The journalist concluded that organic farming is better for biodiversity than conventional farming. Evaluate his conclusion. (5 marks)

e Parts (a) and (b) test your understanding of practical techniques. Part (c) tests your ability to understand results from investigations. Remember that 'evaluate' means 'give the points both for and against' — if you give only one side of the argument, you will not be able to access full marks.

Student A

(a) The number of different species present ✓ and the number of individuals of each species ✓.

e 2/2 marks awarded This achieves both marks.

(b) Use a random number table to obtain random x and y coordinates ✓. Place the bottom left-hand corner of the quadrat at this point ✓. Count the number of individuals of each species in each quadrat ✓. Repeat this at least 10 times.

e 3/3 marks awarded This is a detailed answer that achieves all 3 marks. Any method of placing the quadrats at random would have been allowed, such as dividing the area into a grid and using random numbers to select which squares on the grid to sample.

(c) Most of the studies show biodiversity is better with organic farming ✓, although there are still several studies that show it has no effect or even that conventional farming is better. For example, bees, butterflies and mammals do not seem to be much more successful on organic farms than conventional farms ✓, but some organisms such as plants and beetles are much better ✓. The journalist has reached a conclusion but we don't know if all these studies were done in the same way or at the same time of year ✓. We also don't know if they were done on different kinds of farms. For example, an organic arable farm may be better for biodiversity than a conventional organic farm, but this may not be true for livestock ✓.

e 5/5 marks awarded This is an excellent answer gaining all 5 marks. This student has based the answer on data from the graph, as well as questioning the methodology involved in analysing the data from many different studies. He has also questioned whether this is true for all kinds of farms. There is good reference to the data in the graph.

Student B

(a) The number of different species present ✓.

e 1/2 marks awarded This gains only 1 mark for an incomplete answer.

(b) Place a quadrat frame at random places in the meadow and record the organisms in each quadrat.

e 0/3 marks awarded This is too vague. Student B doesn't tell us how to place the quadrat at random, nor how to record the organisms. There is also no indication of how many quadrats should be used. This is awarded no marks.

(c) Yes, it is better because most of the studies show that there are more species present when organic farming is used ✓. However, there are a few studies that do not show this ✓. Also, organic farming seems to be better for the biodiversity of some groups of organisms than for others, e.g. plants and beetles more than butterflies ✓.

e 3/5 marks awarded This answer gains 3 marks. Student B recognises that the general trend of the data support this, but this is not true for all the studies and the trend is not as marked for some organisms as it is for others. This is backed up by reference to the graph.

Question 7

Table 5 shows the number of differences in the amino acid sequences of cytochrome c between horses and several other organisms.

	Horse	Pigeon	Tuna	Silkmoth	Wheat	Yeast	Bacterium
Horse	0	11	18	27	41	42	64
Pigeon	11	0	17	25	41	41	64
Tuna	18	17	0	30	44	43	65
Silkmoth	27	25	30	0	40	44	65
Wheat	41	41	44	40	0	42	66
Yeast	42	41	43	44	42	0	69
Bacterium	64	64	65	65	66	69	0

Table 5

(a) **Explain what these data tell you about the relationship between these organisms.** (3 marks)

(b) **Scientists use data like these to classify organisms into a phylogenetic hierarchy. Explain the meaning of the words:**

(i) **phylogenetic**

(ii) **hierarchy** (2 marks)

(c) **Another way to investigate relationships within and between species is to examine mRNA and DNA sequences. Suggest and explain one advantage of comparing mRNA sequences rather than the DNA sequences from which the mRNA is transcribed.** (2 marks)

ⓔ This question tests your understanding of evidence for classification, as well as your understanding of a couple of terms. This should not be a demanding question if you understand the topic.

Student A

(a) The more differences in amino acid sequence, the less related that organism is to a horse ✓. The organism that is most closely related to a horse is a pigeon ✓and the least closely related is a bacterium ✓.

ⓔ **3/3 marks awarded** This gains full marks for explaining how data like these are used and then giving examples from the table.

(b) **(i)** Phylogenetic is how closely related, in evolutionary terms, organisms are ✓.

(ii) Hierarchy is large groups subdivided into smaller groups with no overlap ✓.

ⓔ **2/2 marks awarded** This gains both marks for a clear response.

(c) mRNA does not contain introns ✓, but the DNA it is transcribed from contains introns. Therefore, mRNA will give the base sequence that the actual protein is translated from ✓.

e **2/2 marks awarded** This is a valid answer, well explained, gaining both marks.

Student B

(a) The organism on the table that is most closely related to a horse is the pigeon ✓, then a tuna. The least closely related are bacterium ✓and then yeast. The fewer differences in amino acid sequence between an organism and the horse, the more closely related to a horse the organism is ✓.

e **3/3 marks awarded** This also gains full marks. This is a fairly straightforward question.

(b) (i) Phylogenetic means how closely related the organisms are.

(ii) Hierarchy means divided up into groups.

e **0/2 marks awarded** This does not gain any marks. Student B has not explained that phylogenetic refers to evolutionary relationships and a hierarchy involves large groups being divided into smaller, discrete groups.

(c) mRNA is a lot shorter and single-stranded ✓, so it takes less time to find the sequence ✓.

e **2/2 marks awarded** This is acceptable and gets both marks.

Question 8

(a) Describe the processes involved in exhalation in a healthy person. (5 marks)

Figure 6 shows the effect of smoking on lung function.

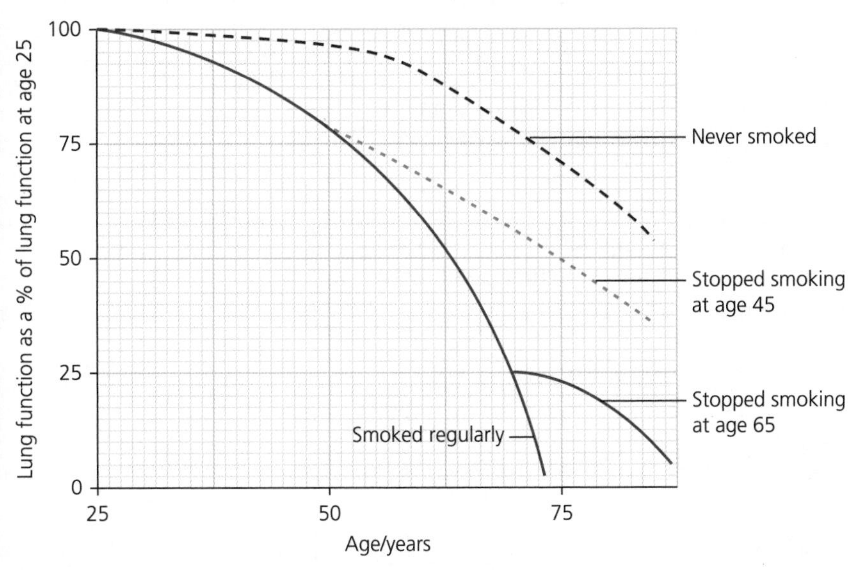

Figure 6

(b) Give two conclusions that can be drawn from the evidence in Figure 6. (2 marks)

Figure 7 shows lung tissue from a person with healthy lungs and a person who has smoked regularly. Both illustrations are drawn at a magnification of ×200.

(a) Healthy lungs

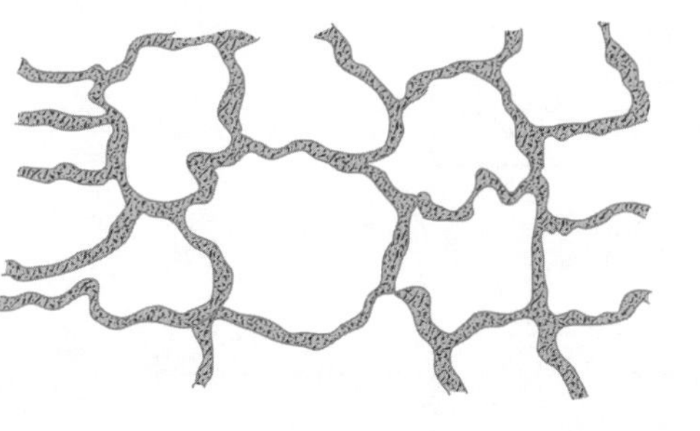

(b) Lung tissue from a regular smoker

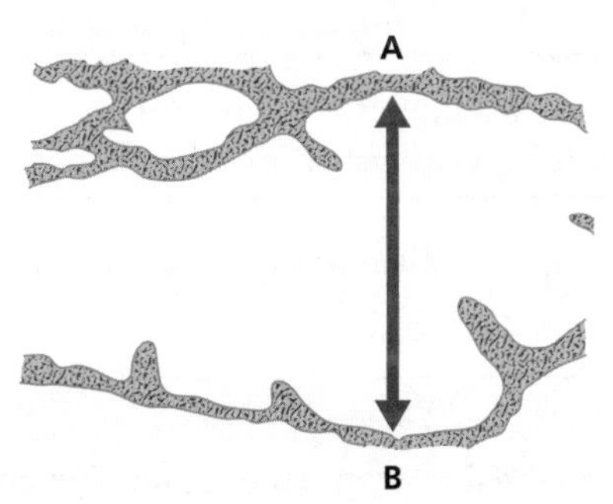

Figure 7

(c) (i) Calculate the actual width of the alveolus along line AB. Show your working. (2 marks)

(ii) The regular smoker had difficulty in breathing. Give two pieces of evidence from Figure 7 to explain why. (4 marks)

e Part (a) is straightforward recall, but remember it is a waste of time writing all about ventilation when the question focuses only on exhalation. Part (b) tests your ability to interpret evidence on a graph. Part (c) tests your ability to calculate magnification and your understanding of lung disease. Note that you must give evidence from the diagram in your answer.

Student A

(a) The intercostal muscles relax ✓, so the rib cage moves down and in ✓. The diaphragm relaxes and returns to its original domed shape ✓. This reduces the volume of the thorax ✓ and increases its pressure ✓, so air is pushed out of the lungs.

e **5/5 marks awarded** This achieves all the marks for correctly naming the muscles and what they do, as well as for detailing the volume and pressure changes in the thorax.

(b) People who smoke at any time have lower lung function than lifetime non-smokers ✓. If you stop smoking at age 65, your lung function is still better at age 75 than people who continue smoking ✓.

e **2/2 marks awarded** This scores 2 marks for two valid points.

(c) (i) AB measures 27 mm = 27 000 µm

$$\text{real size} = \frac{\text{measured size}}{\text{magnification}}$$

$$= \frac{27000}{200}$$

$$= 135\,\mu\text{m}$$

e **2/2 marks awarded** 2 marks awarded for the right answer. The student has shown their working, which means that 1 mark for correct working could be awarded even if there had been a mathematical error.

(ii) The alveoli have broken down, giving fewer larger alveoli ✓ so the surface area for gas exchange is reduced ✓. The walls of the alveoli are thicker ✓, so the diffusion pathway for gas exchange is longer ✓.

e **4/4 marks awarded** This gains all 4 marks for two valid features and explanations.

Student B

(a) The ribs move down and the diaphragm moves up. This makes the thorax smaller and the pressure higher ✓, so air is forced out.

e **1/5 marks awarded** This achieves only 1 mark because a lot of technical detail is missed out. The ribs don't move down on their own — the intercostal muscles relax, allowing them to move down. The diaphragm relaxes, which causes it to move upwards. The thorax decreases in volume, rather than getting 'smaller', but the pressure comment is correct and obtains 1 mark.

(b) If you stop smoking at 65, your lung function worsens less slowly than if you continue smoking ✓. If you give up smoking when you are 45, you have worse lung function than non-smokers but it is much better than if you continue smoking ✓.

e 2/2 marks awarded This scores 2 marks for two valid points.

(c) (i) AB = 27 cm
Real size = 13 µm

e 0/2 marks awarded No marks awarded. The student has used the wrong units (cm instead of mm). No working is shown so no credit can be given for a correct method.

(ii) The alveoli are bigger ✓, which means there is a lower surface area for gas exchange ✓. There is also more mucus in the airways, making it harder to ventilate the lungs.

e 2/4 marks awarded This gains 2 marks for the points on larger alveoli and reduced surface area, but the other feature mentioned (mucus) is not evidenced by the illustration.

Question 9

(a) Describe how proteins are digested in the human gut. (5 marks)

(b) Describe how amino acids are absorbed by co-transport in the small intestine. (5 marks)

(c) Describe how the DNA in the nucleus of a cell is transcribed into mRNA that codes for a protein. (5 marks)

e This is a straightforward question for a student who has revised thoroughly. Although it is the last question on the paper, it may well be the easiest one if you have prepared well for the exam. It might be a good idea to start with this last question in your real exam, to make sure you don't run out of time to answer these straightforward questions fully. Remember that you can answer the questions in any order that suits you.

Student A

(a) In the stomach, an endopeptidase hydrolyses proteins to polypeptides ✓. It breaks peptide bonds within the polypeptide chain forming shorter polypeptides ✓. In the small intestine, an endopeptidase from the pancreas does the same thing ✓. There is also an exopeptidase that hydrolyses dipeptides from the ends of the polypeptides and peptides ✓. In the membrane of the microvilli there is a dipeptidase that hydrolyses dipeptides into amino acids ✓.

e 5/5 marks awarded This scores all 5 marks for a considerable amount of correct detail.

(b) Sodium ions are actively transported out of the cell into the capillary ✓ to maintain a low concentration of sodium ions inside the cell ✓. This creates a concentration gradient for amino acids to be transported into the cell from the gut lumen ✓ alongside a sodium ion ✓. The amino acid is removed from the cell by facilitated diffusion into the capillary ✓.

ⓔ **5/5 marks awarded** This has all the detail required and scores a full 5 marks.

(c) A section of the DNA 'unzips' ✓ and RNA nucleotides line up against one of the strands by complementary base pairing ✓. The enzyme RNA polymerase joins the nucleotides together ✓ to make a molecule of pre-mRNA ✓. Then introns are spliced out to make mRNA ✓. This can enter the cytoplasm to code for a protein.

ⓔ **5/5 marks awarded** This gets all 5 marks. How the DNA splits and the fact that one strand is a template is mentioned. The formation of RNA is described in detail, including the role of RNA polymerase. Finally, the fact that introns are removed from the pre-mRNA to make functional mRNA is mentioned.

Student B

(a) In the stomach a protease hydrolyses proteins to polypeptides. In the small intestine, proteases from the pancreas hydrolyse polypeptides to amino acids ✓ which are absorbed.

ⓔ **1/5 marks awarded** This gains 1 mark for proteins to polypeptides to amino acids, but nothing else. A lot of detail has been missed out.

(b) The amino acid is co-transported into the cell ✓ alongside a sodium ion ✓. The amino acid diffuses into the capillary. Sodium ions are actively transported out of the cell into the capillary ✓.

ⓔ **3/5 marks awarded** This gains 3 marks, but some detail is missing. It is not clear that the amino acid enters the capillary by facilitated diffusion. Also the effect of active transport of sodium ions in creating a low concentration of sodium ions within the cell is not mentioned.

(c) The DNA unwinds and mRNA lines up alongside it by complementary base pairing. Introns are spliced out ✓ and then the mRNA enters the cytoplasm by a nuclear pore.

ⓔ **1/5 marks awarded** This implies that the mRNA is already in the nucleus and lines up alongside the DNA. This is incorrect and gets no credit. However, there is 1 mark for introns being spliced out of the mRNA.

Required practical answers

Required practical 6

1 To stop bacteria from the student or the environment entering the dish; to stop bacteria from the dish or broth culture contaminating the student or environment.

2 If bacteria come from someone, there may be pathogens present even though the person is healthy. If just one pathogen is placed on the Petri dish, after incubation there will be thousands of pathogens present as the bacteria will have replicated many times. There is the risk that these could escape and cause an infection.

3 The student should measure the diameter, radius or area of the inhibition zones. The usual measurement is the diameter, in mm.

4 A or CM because they have produced the largest inhibition zones.

5 The antibiotics might be expensive, and a slightly less effective one much cheaper; the antibiotics might have unpleasant side effects; the antibiotics might be broken down when taken orally; the patient might be pregnant and the antibiotics unsuitable for pregnant women; the antibiotics might have an adverse reaction with other medication that the patient is taking.

Knowledge check answers

1 This increases the surface area available for gas exchange and the capillaries maintain a concentration gradient.

2 a Microvilli increase the surface area.

b Molecules only have a cell-surface membrane to cross to enter the cell.

3 The ends of the tracheoles where they are in contact with the muscle cells.

4 The blood and water would reach equilibrium, so the blood would not become fully oxygenated and the water would leave the gills with a lot more oxygen still dissolved in it.

5 a Many gill filaments; lamellae on gill filaments.

b Capillary wall and lamellae both composed of a single layer of thin, flattened cells.

c Capillaries constantly removing oxygenated blood and bringing deoxygenated blood; counter-current flow between water and blood; ventilation of the gills bringing a constant supply of oxygenated water and removing water that is depleted in oxygen.

6 At night, only respiration is occurring so oxygen diffuses into the mesophyll cells and carbon dioxide diffuses out.

7 This means they can open the stomata at night when it is cooler, and close the stomata in the day when it is hotter and the risk of drying out is greater.

8 a Scale bar measures 9 mm and represents 10 µm.

$$\text{magnification} = \frac{\text{measured size}}{\text{actual size}} = \frac{9000}{10} = \times 900$$

b Various answers can be obtained depending on where the diameter of A is measured. Assuming diameter = 38 mm:

$$\text{actual size} = \frac{\text{measured size}}{\text{magnification}} = \frac{38000}{900} = 42\,\mu\text{m}$$

Diameter of red blood cell measures 9 mm = 9000 µm

$$\text{actual size} = \frac{\text{measured size}}{\text{magnification}} = \frac{9000}{900} = 10\,\mu\text{m}$$

9 $500\,cm^3 \times 15\,min^{-1} = 7500\,cm^3\,min^{-1}$ or $7.5\,dm^3\,min^{-1}$

10 The person will be more tired and become breathless, especially on exertion. This is because the surface area of alveoli is less as the alveoli break down and the fibrous tissue will increase the diffusion distance. The fibrous tissue will also make breathing movements more difficult as the alveoli will lose their elasticity.

11 The hydrogen and ionic bonds holding the protein in its tertiary structure will be broken by the hydrogen ions present and the enzyme will become denatured. Therefore, the active site will change shape and starch will no longer fit into it.

12 Glycosidic.

13 Glucose and galactose are similar in shape and so fit into the same transport protein, whereas fructose is a different shape and will not fit in.

14 Ester bond.

15 This helps the baby to absorb antibodies from the mother.

16 About 25% more.

17

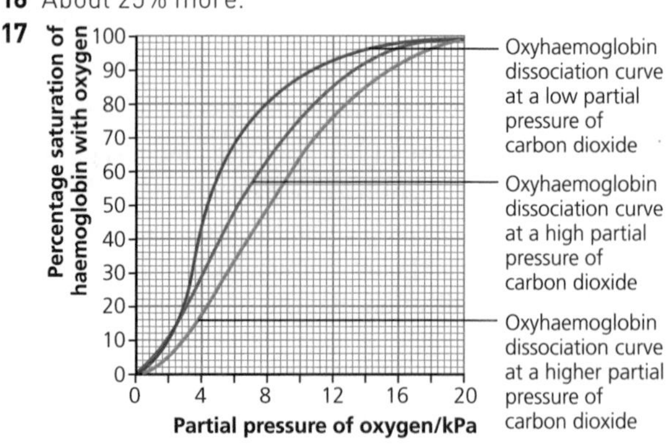

18 It will be to the left — approximately the same shape as curve A.

19 The carbon monoxide binds to the haem groups and therefore they cannot bind to oxygen. In effect, this means there is less haemoglobin in the person's blood. If too much haemoglobin is inactivated in this way, the person will die of asphyxia. Pregnant women should not smoke because smoking produces carbon monoxide, which reduces the oxygen concentration in the mother's blood. In turn, the baby receives less oxygen, which can harm it as a rapidly growing foetus has a high respiration rate.

20 These stop the atrioventricular valves being pushed back too far so that they leak. They ensure the valves can open in one direction only.

21 One cardiac cycle lasts about 0.72 seconds. 60/0.72 = 83 bpm (accept answers in the range 86–82 bpm)

22 It would follow the same pattern as the pressure in the left ventricle, but the line would be lower during systole.

23 The same volume of blood passes through them as flows through the capillaries, but the total cross-sectional area of the veins is lower than that of the capillaries. Therefore, the flow rate must be higher.

24 High blood pressure causes more tissue fluid to form at the arterial end of the capillary, but the amount of water reabsorbed at the venule end — or the amount of fluid returning to the lymph vessels — is the same.

25 As these children have a protein deficiency, their blood does not contain as much plasma protein as it should. Therefore, the water potential of the blood at the venule end of the capillaries is not low enough for sufficient water to be reabsorbed from the tissue fluid.

26 This is because the column of water in the xylem is under greater tension during the day, when transpiration is higher, but the roots may not be able to take in enough water to replace transpiration losses. The cohesion of water molecules means the water column does not break, but 'pulls' the walls of the xylem vessels inwards slightly.

27 There may be air in the xylem from where the flowers have been cut and transported without being in water, although transpiration has occurred. This air will break the water column in the xylem, making it harder for water to travel up the xylem. By cutting off the lower part of the stem underwater, this air gap can be cut off and the water column is not broken as this happens underwater.

28 Measure the distance travelled by the meniscus in a fixed period of time. The volume of water taken up in this time can be found using the formula for the volume of a cylinder ($\pi r^2 d$), where r = radius of capillary tube and d = distance travelled by the meniscus. This volume can be divided by time in minutes to find the volume of water taken up per minute.

29 a Sink **b** Source **c** Sink

30 This ensures you are investigating the effects of removing the phloem and not the effects of a lack of water as well.

31 The second diagram shows organic substances cannot pass up above the ring where phloem is removed. The third figure shows organic substances can travel above the source, as there is no ringing there, but cannot pass down to the roots beyond the point of ringing. The fourth figure shows that organic substances cannot pass either up or down the plant beyond the rings. However, the first figure, where there is no ringing, shows that organic substances can pass both upwards and downwards from the source when there is no ringing. This is a control. This supports the mass flow hypothesis by showing that organic substances travel in the phloem, and can travel both up and down the plant from source to sink. However, this does not verify the exact mechanism by which this happens.

32 Plasmids

33 a UAC **b** AUG

34 Condensation

35 A homologous pair of chromosomes is a pair of chromosomes, one from the mother and one from the father, found in a diploid cell. They share the same genes but not necessarily the same alleles. A pair of chromatids makes up a single chromosome. One chromatid is a copy of the DNA that makes up the chromosome, but it is still attached to the original by the centromere. Once the centromere divides, the chromatids will each become separate chromosomes.

36

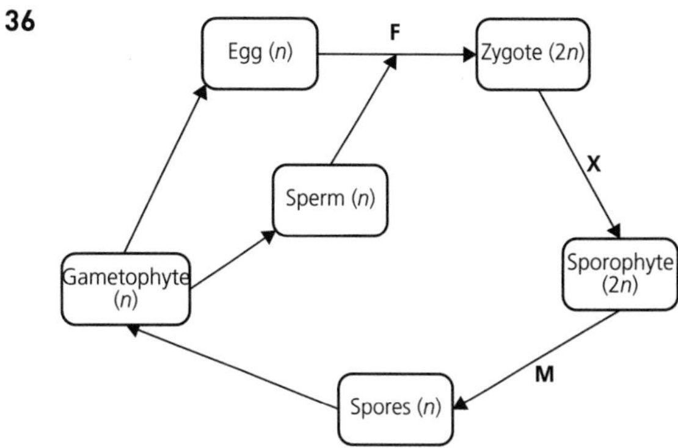

37 These genes pre-date the use of antibiotics in modern medicine but they must have arisen through random mutation. (Note that the presence of antibiotics does not cause antibiotic-resistant genes to arise, but the use of antibiotics creates a selection pressure whereby these genes are advantageous.)

38 a Stabilising

b If the clutch is too small, there may not be enough surviving offspring to replace the parents and ensure survival of the species. If clutch size is too big, it may be impossible for the parents to provide enough food, so most of the offspring die, again making survival of the species less likely.

39 It ensures the other animal is ready for mating; it allows the animal (usually the female) to select a mate with desirable qualities; it triggers hormonal changes that are necessary for egg-laying.

40 *Australopithecus africanus* is less closely related to either of the *Homo* species than the two *Homo* species are to each other.

41 Those with fewer differences from humans are most closely related to humans, so in order of closeness of relationship to humans, the order is chimpanzee, baboon and dog.

42 a Species richness = 6

Species diversity $d = \dfrac{N(N-1)}{\sum n(n-1)}$

$$= \frac{25(24)}{3(2) + 4(3) + 7(6) + 2(1) + 4(3) + 5(4)}$$

$$= \frac{25(24)}{6 + 12 + 42 + 2 + 12 + 20} = \frac{600}{94} = 6.38$$

b Species richness only takes account of the number of species present, whereas species diversity also considers how many of each species are present, so this may be more sophisticated. However, species diversity indices are only meaningful when compared. An area with very few species but a lot of each species can have a higher species diversity index than an area with many species but few of each species.

43 This shows that macaques are the most genetically related to humans, followed by dogs, then birds, then frogs and finally lampreys.

Index

Index